MANUAL OF DISEASE AND INSECT-PESTS OF IMPORTANT VEGETABLES

Dr. Satish K. Gupta
Professor (Plant Pathology)
Shoolini University of Biotechnology and
Management Sciences, SOLAN - 173 212 (H.P.)

Dr. K. C. Sharma
Principal Scientist (Retd.)
Department of Entomology,
Dr Y S Parmar University of
Horticulture and Forestry,
Nauni, SOLAN - 173 213 (H.P.)

SCIENTIFIC®
PUBLISHERS

Published by:

SCIENTIFIC PUBLISHERS (INDIA)
5A, New Pali Road, P.O. Box 91
Jodhpur 342 001, India
E-mail: info@scientificpub.com
Website: www.scientificpub.com

ISBN: 978-93-89412-28-4
eISBN: 978-93-89412-30-7

Printed in India

Shoolini University

Shoolini University of Biotechnology & Management Sciences

Post Box No. 9, Head Post Office, The Mall, Solan-173212, H.P.
Saroj Villa, Officers Colony, Kotla Nala, Rajgarh Road, Solan-173212, H.P.
Tel: Off. (01792) 226674, 308000 • Fax: (01792) 308000
Email: vc@shooliniuniversity.com • khoslapk_2001@rediffmail.com

FOREWARD

Dr. P.K. Khosla
Vice Chancellor

Agriculture has always been central to India's economic growth. The sector has provided much cushion to the economy in times of economic turbulence, including the present in the testing times of Corona pandemic. Emphasis on scientific agriculture has been instrumental in taking the country from the stage of food deficient to food surplus. It is established that with economic growth the share of agriculture in GDP of a country would go down; because of the high value nature of the products of industrial and services sectors. But agriculture still contributes about 18% to our GDP and accounts for nearly 55% of the total employment. By virtue of its diverse climatic conditions India is in a position to produce an array of crops like cereals, millets, pulses, vegetables, flower and fruits.

Economic development of the country has also brought in the change in our consumption habits. Higher per capita income and improved food processing and supply chains have made it possible for the average Indian consumer to diversify his/her consumption basket. This diversification of consumer goods also includes varied traditional and exotic agricultural crops, especially fruits and vegetables. With our attaining the self sufficiency in production of foodgrains the emphasis now is on high value and nutraceutical crops. As result, the area under horticulture crops has reached a high of 25.87 million hectares at present and out of which 10.43 million hectares is under vegetables. A look at the recent growth rates of vegetable production suggests that scientific vegetable production is paying its dividend. Along with high returns to growers, vegetable cultivation also generates employment by way of practices like packing, forwarding, processing and marketing. With our attaining self sufficiency in production of foodgrains, emphasis now is on high value fruits and vegetables and nutraceutical crops.

However, vegetables are highly perishable and get infected/ damaged by various diseases and insect-pests. Estimated 15-20 percent losses, both at pre and post-harvest stages, are attributed to diseases and insects-pests.The ravages of these diseases and insect-pests can be kept in check by adopting timely and effective management practices. Formulating management strategies for a particular disease/ insect-pest requires comprehensive knowledge of diagnosis/ damage, epidemiology/ biology of the pathogen/ insect as well as the host. Conventional plant disease control based on chemical pesticides alone may not be acceptable in the present scenario, where protection of the environment and human health is a serious consideration.

Students, both at undergraduate and postgraduate level, show interest and acquire better understanding of a subject if it is presented in an easy to understand manner. I am happy to know that a manual entitled ***"Diseases and Insect-pests of Important Vegetables"*** is being published by the School of Agriculture and sincere efforts have been made by the authors to describe the diseases and insect-pests of major vegetables in a simplified manner. Thesymptomatology, pathogen, predisposing factors and identification and damage of eachdisease and insect-pest, along with management through host resistance, cultural, biological and chemical methods have been described in the manual. I am sure that this manual will not only serve the needs of students, but will also be helpful to extension workers and the elite growers.

I appreciate the efforts made by the authors in publishing this manual.

(P.K. Khosla)

PREFACE

Vegetables constitute an integral part of our daily diet, providing minerals, vitamins and essential fibers. India is the second largest producer of vegetables globally, after China, but our productivity is relatively low. Over time, many high yielding varieties and hybrids of various vegetable crops have been developed which not only increased the production but also the quality of the produce. Nevertheless, various diseases and insect-pests are proving major impediments in successful production of these crops. Further, some diseases and insect-pests, which were of minor importance, have become major ones due to climate change. Introduction of new diseases and pathogens and insect pests through import of plant material and germplasm is another area of concern.

Proper diagnosis of diseases and insect-pests is thus a very important management aspect for reducing the losses. Without proper knowledge, farmers are applying synthetic chemicals indiscriminately to manage the diseases and insect-pests. This not only results in environmental pollution and health hazards but also affect the quality of the produce to meet the international sanitary and phytosanitary standards.

The present Manual, entitled ***"Diseases and Insect-pests of Important Vegetables"*** is an effort to compile available information on various diseases and insect-pests with regard to symptomatology, pathogens, predisposing factors and identification and damage of diseases and insectpests, along with management through host resistance, cultural, biological and chemical methods. The Manual contains a total of sixteen chapters, with illustrations and coloured photographs, dealing with diseases and insect-pests of important vegetables. Besides this, information on various aspects of pesticides, their common formulations and other related information on safe handling of pesticides have also been included. This will help the stakeholders to decide when and which pesticide application is to be done. It is hoped that the information contained in this manual will be useful to a wider section of stakeholders including the students, teachers, researchers and extension workers engaged in agriculture and horticulture and the elite growers.

Authors are thankful to Dr. P.K. Khosla, Honb'le Vice Chancellor, Professor Atul Khosla, Pro-Vice Chancellor and Dr. Y.S. Negi, Dean, School of Agriculture, Shoolini University, Solan for their continuous encouragement and providing necessary facilities to write this Manual. Authors are also grateful to Dr. P.K. Chakrabarty, Member, Agricultural Scientists Recruitment Board and Dr. Monica Sharma, Scientist, Plant Pathology, Dr. Y.S. Parmar University of Horticulture and Forestry, College of Horticulture and Forestry, Neri for their suggestions provided for improvement of the manuscript. We are also thankful to Dr. Neeraj Pizzar, Chairman, Publication Cell of the University for the improvement of language and subject matter of the manuscript.

S. K. Gupta
K. C. Sharma

ABOUT AUTHOR

Dr. S.K. Gupta, working as Professor (Plant Pathology), School of Agriculture, Shoolini University of Biotechnology and Management, Solan and former Professor and Head, Department of Plant Pathology, Dr. Y.S. Parmar University of Horticulture and Forestry, Nauni, Solan (H.P.). He did his M.Sc. and Ph.D. from HPKV, Palampur and obtained advanced training on Apple Scab Monitoring Warning from Sudtiroler Beratungsring Fur obst-und-Weinbau, Lana, Italy. Dr. Gupta worked in various research projects sponsored by ICAR, NATP, Mini Mission (ICAR) NAIP and HIMCOSTE. He developed E-contents for the B.Sc. (Hort.) course entitled "Diseases of Vegetables, Ornamental and Spice Crops. He has been working on epidemiology and Management of Horticultural crop diseases since last 35 years and guided 11 M.Sc. and 3 Ph.D. students. He is recipient of many awards including PPSinghal Pesticide India Award thrice (1998, 2002, 2003) and prestigious Sharda Lele Memorial Award, 2019 of Indian Phytopathological Society, New Delhi. Dr. Gupta was elected as Zonal President (NZ) of Indian Phytopathological Society for the year 2012 and Vice President of Indian Society of Plant Pathologists in 2017. He has published 105 research paper in the Journals of National and International repute, 22 review/ book chapters, one Manual, four Research information Bulletins and five Books. Dr. Gupta visited Italy and France in connection with various academic pursuits.

Dr. K.C. Sharma retired as Principal Scientist from the Department of Entomology, Dr YS Parmar University of Horticulture and Forestry, Nauni, Solan. He did his M.Sc. from Himachal Pradesh Agriculture University, Palampur and Ph.D from Dr YS Parmar University of Horticulture and Forestry, Nauni, Solan. He has more than 27 years of experience of teaching various courses in Entomology to B.Sc., M.Sc and Ph.D students. He has also a vast experience in handling various ad hoc research projects sponsored by the ICAR, DST, UGC, MOE&F and Mini Mission. Besides, he had been the Co-PI of many research projects sponsored by various funding agencies. He has guided 12 post graduate students for the last 30 years and has published 101 research papers in the journal of national and international repute and three manuals. He has been awarded the appreciation letter by the University for his contribution in the field of Entomology.

CONTENTS

Foreword *iii*

Preface *v*

1. **Introduction** 1
2. **Nursery Diseases and Insect-Pests** 3
3. **Diseases and Insect-Pests of Tomato** 8
4. **Diseases and Insect Pests of Bell Pepper and Chili** 29
5. **Diseases and Insect- Pests of Brinjal** 42
6. **Diseases and Insect-Pests of Cucurbits** 52
7. **Diseases and Insect-Pests of French Bean** 61
8. **Diseases and Insect- Pests of Okra** 75
9. **Diseases and Insect- Pests of Pea** 83
10. **Diseases and Insect-Pests of Crucifers** 96
11. **Diseases and Insect-Pests of Carrot** 113
12. **Diseases and Insect- Pests of Onion and Garlic** 121
13. **Diseases and Insect-Pests of Colocasia** 130
14. **Fungicides in Vegetable Disease Management** 133
15. **List of Pesticides which are Banned, Refused Registration and Restricted in use** 139
16. **Safe Handling of Pesticides** 188

Selected References 190

Glossary 192

1

INTRODUCTION

Ever increasing human population warrants for sustainable food production. Among different food items such as cereals, pulses, oil seeds, fruits etc. vegetables provide maximum nutrient sources and are short duration crops. Vegetables are considered as "Protective Supplementary Food" as they are rich in minerals, vitamins and amino acids which are required for normal functioning of human metabolic processes. Thus vegetables constitute an important part of balanced diet and play a vital role in the maintenance of good health.

India is endowed with the varied climatic conditions, which are suitable for production of various vegetables round the year in one part or the other. Before, 1960's, mostly traditional crops were grown and vegetables were grown on limited areas. Recently the area under vegetables has increased tremendously and now India is the second largest producer of the vegetables after China. Different vegetables like tomato, capsicum, beans, cucumber, peas, cauliflower, cabbage, radish, turnip, carrot, garlic, onion, colocasia, okra etc. are grown on large area to meet the demand of ever increasing population of the country.

During cultivation and at postharvest stages, these vegetables are attacked by several diseases and insectpests, which have become major constraints in the production of these vegetables as they not only reduce the yield but also impair the quality of the produce. In view of the lucrative returns from the vegetables being cash crops, farmers with limited land holdings are forced to follow monoculture and intensive cultivation, which also ensures the perpetuation of various pathogens and insect-pests year after year. Under such conditions, management of diseases and insect-pests damaging vegetables both under field and at post harvest stages cannot be ignored.

Much awareness about the importance of plant diseases was created after the epidemics of late blight of potato occurred in 1840's in Ireland. Similarly several insect-pests of vegetables have also appeared in epidemic form in recent past and caused substantial yield losses. Some insects like, whitefly, aphids, grasshoppers etc. are also acting as vectors of various virus diseases and ensure their appearance year after year.

The efficient control of different diseases and insect-pests is possible only through adequate knowledge of crop husbandry, physical and chemical properties of soil, symptoms/ infestation of disease/ insect-pests, survival and dissemination and effect of the environmental conditions on the disease/ insect-pests development and their progress. The information on different diseases and insect-pests affecting vegetable crops has been included in this manual. The symptoms of different vegetable diseases, identification and damage of insect-pests infecting various vegetables are described in detail in this manual alongwith predisposing factors and management with cultural, biological, plant origin pesticides, chemicals and integrated pest management have been described in this manual.

✦

2

NURSERY DISEASES AND INSECT-PESTS

The nurseries and young seedlings of vegetable crops like tomato, bellpepper, chillies, brinjal, cucumbers, cabbage, cauliflower, broccoli etc. for which nursery is raised for transplanting are affected by several diseases and insect pests which substantially impair the quality of the nursery. They are also responsible for the poor seed germination and seedling stand but also for carryover of the pathogens/insect-pests to the field where transplanting is done for raising of the crop.

DISEASES

a) Fungal Diseases

Damping-off

The disease is caused by species of *Pythium, Phytophthora, Fusarium, Rhizoctonia solani, Phoma* spp., *Phomopsis vexans* etc. in different crops. Out of these, species of *Pythium* are more frequently observed with this disease.

In vegetable nurseries of all those crops for which we raise seedlings for transplanting, it is one of the major diseases due to which seedlings are either killed before emergence or after emergence. The disease is prevalent in poorly managed nursery beds, incited by the fungi which are seed or soil borne in nature.

Symptoms

The disease manifests in following two phases:

Pre-emergence damping-off

In nursery beds when seedlings fail to emerge from the soil either due to seed rots or killing of young seedlings before their emergence from soil, hence resulting in patchy appearance of seedlings stand in early stages of growth (Plate 1a).

Post emergence damping-off

This phase is characterized by toppling over of infected seedlings at any time after their emergence from the soil. The infected tissue initially appears to be water-soaked and soft. Subsequently, the stem at the infection point gets constricted resulting in their mortality. In the nurseries, the disease usually radiates from initial infection points and thus gives the appearance of islands wherein almost all seedlings are killed (Plate 1b).

Pathogen(s)

Species of *Pythium, Phytophthora, Fusarium, Rhizoctonia solani, Phoma* spp., *Phomopsis vexans* etc. are responsible for this disease. Out of these, species of *Pythium* are more frequently observed with this disease. Among *Pythium* spp., *P. aphanidermatum* is commonly observed. The *P. aphanidermatum* belongs to the kingdom Chromista, division Eumycota, phylum Oomycota, class Oomycetes, order Peronosporales and family Pythiaceae. The mycelium is coenocytic. The sporangia are lobbed, toruloid and branched. Oogonia are spherical, smooth walled, terminal on lateral hyphae. Antheridia are monoclinous, intercalary or terminal. Oospores are aplerotic, single and moderately thick walled.

Predisposing factors

The pathogens associated with this disease are either seed or soil borne or both. High temperature, soil drying, excessive salt concentration and water logging may predispose the seed or seedling to the disease. The disease is further aggravated in ill-aerated soils with poor drainage having thick stand of the seedlings.

Management

1. Change the nursery site every year.
2. Either solarize the soil with transparent polyethylene (25µm) sheet for 45 days during summer months or treat the soil with Formalin (5%) atleast 21 days before sowing or apply bio-agents like *Trichoderma harzianum/ T.viride* (40g/ m^2).

3. Use healthy seed and treat it with captan (0.3%).
4. After seedling emergence from the soil, drench the bed with captan (0.3%) alone or in combination of carbendazim (0.1%) and repeat at 10 days interval.
5. Give light but frequent irrigations.

b) Bacterial Diseases

Many pathogens of bacterial nature infest the foliage of seedlings in nursery beds after their emergence from the soil surface. Sometimes black rot (*Xanthomonas campestris* pv. *campestris*) in crucifers, bacterial spot (*X. campestris* pv. *vesicatoria*) and bacterial canker (*Clavibacter michiganensis* pv. *michiganensis*) in chilli and tomato etc. appear in nurseries and spread rapidly. Keeping this in view, suitable seed treatments and foliar sprays should be applied as suggested in the subsequent chapters of this manual.

c) Viral Diseases

Many viral diseases like leaf curl and mosaics appear in the seedlings of some vegetables and are carried from nursery to the fields. Most of these diseases are transmitted through their insect vectors like aphids and white fly.

To protect the nursery from these diseases, it is better to raise the nursery under a net house (40 mesh size) so that the nursery is not exposed to the viruliferous vectors. Insecticides should be applied at regular intervals to keep the vectors under check. Any available insecticide dust should be applied around the borders of the bed to restrict the entry of ants and other soil borne insects in the nursery bed.

d) Nematode Disease(s)

If the seedlings of tomato, brinjal and cucurbits are raised in untreated soils, are commonly infested with nematodes like *Meloidogyne* spp. Infested seedling if transplanted in fields may carry the nematodes to the fields where the problem does not exist earlier and can result in severe crop yield losses in later stages. Treat nursery bed soil with Formalin (5%) atleast 21 days before sowing of seed in nursery bed or the management methods suggested in subsequent chapters in this manual may be followed.

(B) INSECT- PESTS

1. White grub (*Anomola* spp., *Holotrichia* spp.) (Coleoptera: Scarabaeidae)

Identification

Adults are variously coloured varying from black, green, brown and dark brown in colour. Larvae are creamy white with stout body. These are 'C'-shaped in structure and always remain in soil.

Damage

Although damage is caused by both adults and larvae (grubs) yet the injury to vegetable seedlings occurs as a result of feeding by the grubs in soil. They disturb the root system. Older second instar and third instar grubs are more damaging.

Management

1. Use of light traps to catch adults and subsequently killing them.
2. Change nursery site every year.
3. Drench nursery beds with chlorpyriphos 20%EC @ 3-4 ml/lt of water.

2. Aphids (*Aphis* spp.) (Hemiptera: Aphididae)

Identification

Various species of aphids are found in the nursery stages. The aphids can be identified by the presence of cauda and cornicles on their body.

Damage

Aphids have piercing and sucking types of mouth parts so damage is done by piercing their stylet in the leaf tissue and sucking of cell sap due to which the leaves start curling and vitality of plant is reduced.

Management

1. Install yellow sticky trap @12 traps/ha to monitor winged adult or Moerick yellow water traps to catch winged form.

2. When this aphid appears, a spray of quinalphos 25%EC (0.025%) may be given. The activity of this pest remains low due to low temperature during winter, however, during February, when the temperature rises again a spray of the same insecticide may be given and if the attack persists,it may be repeated after 15 days.
3. Application of commercially available neem formulations (2–3 ml/litres of water) or neem seed kernel extract @4% was also found effective in reducing the population of this pest.

Besides, some other insect pests like cut worm, whitefly, termites, leafhoppers, grasshoppers, some defoliators, etc. also attack the nursery plants at various stages. Non insect pests (nematodes, mites) and vertebrate pests (squirrels, rabbit, deer, rats) etc. also attack the nurseries of various vegetable crops.

The details of some of these have been given under the insect pests of specific crops.

Disclaimer

Though utmost care was taken in recommending only approved pesticides for management of different diseases and insect-pests, pesticide recommended wrongly if any, may be purely unintended, inadvertent and by oversight. Further no pesticide can be used in commercial scale without CIBRC approved label claim on the target crop.

3

DISEASES AND INSECT-PESTS OF TOMATO

Tomato (*Solanum lycopersicon*), a member of Solanaceae family is one of the most popular crops which is grown globally and consumed as salad, ketchups, sauce, soup, pickles etc. The crop is grown from temperate to sub-tropical regions of the country all the year round. High temperature followed by high humidity during the growth not only favours the luxuriant growth of the crop but also favours the attack of various biotic and abiotic agents like diseases, nutritional disorder and insect-pests which not only reduce the yield but also reduce the market value of the produce. In this chapter important diseases and insect-pests are discussed herewith along with their management practices.

(A) DISEASES

a) Fungal diseases

1. Buckeye rot (*Phytophthora nicotianae* var. *parasitica*)

Symptoms

Spots on tomato fruits are marked with pale brown, concentric rings resembling slightly the markings on a buckeye (Plate 1a). Young and green fruits are more susceptible as compare to mature red fruits. The spots may be small or cover a major portion of the fruit. Later, infected fruits may drop down on the ground (Plate 1b).

Pathogen

The fungus responsible for this disease is *Phytophthora nicotianae* var. *parasitica*. The pathogen belongs to kingdom Chromista, phylum Oomycota, class Oomycetes, order Peronosporales and family Pythiaceae. The mycelium of the pathogen is hyaline and coenocytic with branching typically at right angles. The sporangiophores arise from hyphal threads and produce sporangia. The sporangia are broadly ovoid to globose in shape having one hemispherical papilla at the tip. Chlamydospores are smooth, globose, and slightly yellowish with thick brown walls, produced abundantly in culture and germinate by producing zoospores or germtubes. Antheridia are amphigynous, spherical or oval and oogonia are rough, thick walled and yellowish brown in colour. Oospores are aplerotic, 18 to 20 μm in diameter with 2 μm thick wall.

Predisposing factors

The pathogen perpetuate in the soil either as oospore or chlamydospores. Infection takes place through rains splashed zoospores. The disease assumes serious proportions when moisture or rainfall is plentiful coupled with temperature ranging from 22.5–25° C.

Management

1. Stake the plants erect and remove foliage and fruit up to a height of 15–20 cm.
2. Collect and destruct the infected fruits regularly.
3. To avoid spattering rains, apply grass or pine needle mulch on the field floor.
4. With the onset of monsoon rains, spray the crop with cymoxanil + mancozeb (0.25%) followed by sprays of either mancozeb (0.25%) or propineb (0.3%) or copper oxychloride (0.3%) or Bordeaux mixture (4:4:50) at 7–10 days intervals.

2. Late blight (*Phytophthora infestans*)

Symptoms

Symptoms of the disease appear as brown dead areas on leaves, which initially appear as faded green patches and later turn into brownish black lesions spreading rapidly during wet weather (Plate 2a). Within 3–4 days of wet weather whole plant may get blighted. On tomato fruits, dark olivaceous, greasy–appearing spots are formed (Plate 2b).

Pathogen

The disease is caused by *Phytophthora infestans.* The pathogen belongs to kingdom Chromista, phylum Oomycota, class Oomycetes, order Peronosporales and family Pythiaceae. The mycelium is coenocytic and endophytic. Sporangiophores are slender, hyaline, sympodially branched, indeterminate, relatively thick walled, septate having side branches with swollen base. Sporangia are hyaline, thin walled, lemon shaped, distinctly papillate. On maturity, the sporangium turns to the side, growth of the sporangiophore continues which looks to have zigzag growth with characteristics swellings at the nodes. *P. infestans* is heterothallic fungus and involves two mating types namely A 1 and A 2. In India, although A 1 mating type is of common occurrence, prevalence of A 2 type has been recorded in Shimla Hills after 1984. Anthredia and oogonia are produced after A 1 and A 2 mating types come in close contact. Anthredia anre amphygynous while oogonia are spherical. Oospores are thick walled and develop after fertilization.

Predisposing factors

Prevalence of cool (12 -15°C) and wet weather is conducive for disease development and its rapid spread.The pathogen also infect potato. If the disease first appears on potato, it can also cause infection to tomato or vice-versa.

Management

1. Follow crop rotation and avoid potato in rotation.
2. Improve drainage of the field and keep field free from weeds.
3. Collect and destroy infected fruits regularly.
4. Spray the crop with mancozeb (0.25%) as a protective spray followed by cymoxanil + mancozeb (0.25%)at a critical stage of disease appearance followed by mancozeb or copper oxychloride (0.3%) or Bordeaux mixture (4:4:50) spray at 7 days interval.

3. Alternaria Leaf spots (*Alternaria solani, A. alternata, A. a.* f.sp. *lycopersici*)

Three species of *Alternaria* have been found associated with this disease which are described below:

Symptoms

Alternaria solani: On leaves, dark brown spots develop which increase in size by forming concentric rings that give the spots a target board like effect (Plate 3a). In humid weather, the affected areas coalesce and form dark brown patches.

A. alternata: Small, circular, scattered and dark brown spots appear on leaves (Plate 3b). Older spots are surrounded by yellow halo.

A. a. f.sp. *lycopersici:* Small, angular, scattered and light brown spots appear on the leaves (Plate 3c). Spots are not surrounded by yellow halo. Similar spots also appear on stem (Plate 3d).

Pathogen(s)

Three different species of *Alternaria* are responsible for this disease. The pathogen (s) belongs to kingdom Fungi, phylum Ascomycota, class Deuteromycetes, order Moniliales and family Dematiaceae

Alternaria solani: The mycelium consists of septate, branched, light brown hyphae, which become darker with age. Conidiophores emerge through the stomata from the dead centers of the spot. Conidia are beaked, muriform, dark and borne singly or in chains of two. Five to ten transverse septa and a few longitudinal septa are present in each conidium.

A.alternata: The conidia occurring on leaflet lesions under field conditions are 1-6 septate. The size of beak range from 3.0 - 10.0 μm with an average length of 7.9 μm.

A. alternata f.sp. *lycopersici:* The conidia occurring on leaflet and stem lesions under field conditions are 1–5 septate measuring 17.0 - 50.0 × 10.0 - 13.0 μm (av. 34.40 × 12.30 μm) and beak length ranged from 3.0 - 7.0 μm with an average of 6.13 μm.

Predisposing factors

Alternaria species overwinter in diseased plants debris and infected seed. Temperature ranging from 25–30° C during the cropping season along with humid conditions favours disease development.

Management

1. Collect and destroy diseased plant debris.
2. Follow crop rotation.
3. Field should be weed free.
4. Use healthy seed.
5. Treat the seed with captan (0.3%).
6. Remove foliage up to a height of 15-20 cm to avoid moist and stagnant air conditions.
7. With the initiation of the disease, apply copper oxychloride (0.3%) or mancozeb (0.25%) or captan (0.3%) spray. Repeat sprays at 10 -14 days interval.

4. Septoria leaf spot (*Septoria lycopersici*)

Symptoms

Small, water soaked, more or less circular spots appear on the leaves with brown margins and grey centre (Plate 4).

Pathogen

The disease is caused by *Septoria lycopersici*. The pathogen belongs to kingdom Fungi, phylum Ascomycota, class Deutromycetes, order Sphaeropsidales and family Sphaeropsidaceae. On the host, young mycelium is hyaline, thin walled and sparingly septate. Pycnidia are subglobose, composed of 2-3 layers of brown cells and range between 53-165 µm in width. Pycniospores are filiform, slightly curved, hyaline and septate with pointed or rounded ends. No sexual stage has been reported.

Predisposing factors

The fungus perpetuate in infected plant debris. Temperature of 20- 25° C and 75-92 per cent relative humidity along with muggy weather for 2 - 3 days is congenial for disease development.

Management

1. Destruction of infected plant material and use of clean seed is useful in checking the disease.
2. Follow crop rotation and ploughing under all infected plant material.

3. Use resistant cultivars, if available.
4. Spray the crop with mancozeb (0.25%) or copper oxychloride (0.3%) or thiophanate methyl (0.1%) and repeat at 10–14 days interval.

5. Powdery mildew
(*Leveillula taurica, Erysiphe cichoracearum*)

Symptoms

L. taurica: The symptoms of the disease appear as white talcum like covering on the under surface of the leaves while yellow spots appear on the corresponding upper surface. Premature defoliation of infected leaves occur.

E. cichoracrearum: White powdery lesions appear on all above ground parts of the plant which are initially discrete but under favourable weather conditions gradually enlarge and coalesce (Plate -5). Affected foliage becomes chlorotic, then brown and dies.

Pathogen(s)

Two fungus i.e. *Leveillula taurica* and *Erysiphe cichoracearum* are responsible for this disease. The pathogen belongs to kingdom Fungi, phylum Ascomycota, class Ascomycetes, order Erysiphales and family Erysiphaceae.

L. taurica: The pathogen is described in Bell pepper diseases.

E. cichoracearum: The mycelium is superficial. The conidiophores are hyaline, thin walled and bear chains of conidia which are also hyaline and thin walled and are oblong. Perithecia are dark coloured, spherical with myceloid appendages. Each perithecium contains numerous asci and each ascus contains 2–3 ascospores.

Predisposing factors

Dry weather conditions coupled with low humidity favour *L. taurica* infection while *E. cichoracearum* is favoured by temperatures of 18–24° C and RH ranging from 50–70 per cent.

Management

1. Apply sprays of wettable sulphur (0.2%) and repeat at 10–14

days intervals.Sprays used for the control of other fungal diseases also provide control of this disease to some extent.

6. Fusarium wilt (*Fusarium oxysporum* f.sp. *lycopersici*)

Symptoms

Initial symptoms on young plants appear as clearing of vein lets and drooping of the petioles. Lower leaves of infected plants first show yellowing and with the progress of the disease they die (Plate 6). Later whole plant wilts. Browning of vascular bundles takes place and it can be seen in a cross section of the lower stem.

Pathogen

The disease is caused by *Fusarium oxysporum* f.sp. *lycopersici.* The pathogen belongs to kingdom Fungi, phylum Ascomycota, class Deutromycetes, order Moniliales and family Tuberculariaceae. The mycelium is hyaline at first becoming cream coloured with age and has branches and septations and some isolates produce blue or red pigments. The micro conidia are ellipsoidal, straight to curved, one to two celled and hyaline. Macro conidia are hyaline, 3-5 septate, falcate and appear on sporodochia or without formation of such fruiting bodies. Chlamydospores both rough and smooth walled are produced often solitary but occasionally in pairs or chains.

Predisposing factors

The pathogen is soil borne in nature and once introduced in a field it becomes difficult to eradicate. Hot dry weather favours wilt development. Presence of root knot nematodes and monoculture enhance the disease.

Management

1. Long crop rotation should be followed.
2. Collect and destroy the infected plant debris.
3. Apply Nitrate form of nitrogen.
4. Use resistant varieties like Marglobe or Rutger/ hybrid if available.
5. Treat seed either with bioagents like *Trichoderma viride* or *Streptomyces griseus* or *Bacillus* subtilis before sowing.

6. Remove wilted plants from the field and drench remaining plants with mancozeb (0.25%).
7. Fields having the history of nematode infestation, after 3 days of transplanting drench plants with Velamprime (Fluopyram, 500 ml/ha) or apply Nimitz (Fluensulfone 2% GR, 1g/plant).

b) Bacterial diseases

7. Bacterial spot (*Xanthomonas campestris* pv. *vesicatoria*)

Symptoms

On leaves and stems, water soaked spots appear, which are darker in colour and not surrounded by halo(Plate 7a). Later blightening of foliage takes place that gives a scorched appearance. Fruit spots appear as brownish-black elevated scab like areas with feathered or irregular margins, which later become sunken and appear grey or bleached (Plate 7b).

Pathogen

The bacterium responsible for this disease is *Xanthomonas campestris* p.v. *vesicatoria*. The bacterium belongs to Kingdom Prokaryotae, division Gracilicutes, class Proteobacteria and family Pseudomonadeaceae. The bacterium is rod shaped, Gram negative, strictly aerobic, motile by a single polar flagellum, forming yellow colonies, liquefies gelatine, attacks starch and digests protein.

Predisposing factors

The pathogen is mainly seed borne which give rise to infected seedlings and also overwinters in plant debris. Moderate temperatures (20-25° C) and high relative humidity (100%) favour disease development.

Management

1. Rotate fields to avoid carry over of the bacterium.
2. Collect and destroy infected fruits and plant debris.
3. Collect healthy seed from disease free fruits and treat it by dipping in Streptocycline (100 ppm) solution for 30 minutes.

4. With the initiation of the disease, spray the crop with Streptocycline (100ppm) immediately followed by sprays of copper oxychloride (0.3%) or Bordeaux mixture (4:4:50) and repeat at 7-10 days interval.

8. Bacterial wilt (*Ralstronia solanacearum*)

Symptoms

Initial symptoms of the disease appear as drooping of leaves, without yellowing along with rotting of the stem. The pathogen mostly remains confined to the vascular system (Plate 8) where it produces polysaccharides, which mechanically block the system, and checks the translocation of water resulting in wilting of the plants.

Pathogen

The disease is caused by *Ralstonia solanacearum*. The bacterium belongs to Kingdom Prokaryotae, division Gracilicutes, class Proteobacteria and family Pseudomonadeaceae. The bacterium is Gram negative, rod shaped, frequently occurs in pairs, motile with 1 to 4 polar flagella. It is aerobic and catalase and oxidase-positive and forms nitrites from nitrates.

Predisposing factors

The associated bacterium is soil borne. High soil temperature and plentiful supply of moisture favour disease development.

Management

1. Crop rotation with non-solanaceous crops should be followed.
2. Before transplanting, solarize field soil for at least 45 days.
3. Apply bacterial antagonists like *Pseudomonas fluorescens, P. glumae, P.cepecia* and *Bacillus* spp.
4. Use resistant hybrids like Sun seed 7711.

9. Bacterial Canker (*Clavibacter michiganensis* subsp. *michiganensis*)

Symptoms

Infected plants show wilting of lower leaves. Brown streaks and canker develop on the stem (Plate 9a). On fruits, small brown, scabby lesions develop which are surrounded by white halo that resembles the bird's eye (Plate 9b).

Pathogen

The disease is caused by bacterium *Clavibacter michiganensis* subsp. *michiganensis.* The bacterium belongs to Kingdom Prokaryotae, division Fermicutes, class Thallobacteria and family unknown.The bacterium is aerobic, Gram -positive, coryneform rod with cells growing either singly or in pairs. Pigmented mutants develop due to possession of a range of carotenoid compounds. The organism is characterized by oxidation of carbohydrates, being nonlipolytic, it can only liquefy gelatine slowly, the ability to hydrolyse starch is weak or absent and amino acids like biotin, nicotinic acid and thiamine are required for growth.

Predisposing factors

The pathogen overwinters in infected seed, plant debris and weed hosts from one season to another. Optimum temperature for disease development is 28° C. The disease is more severe in wet weather.

Management

1. Infected plant debris should be collected and destroyed.
2. Do not grow tomato in those fields which have disease history for at least 3 years and follow crop rotation.
3. Use disease free seed and treat it with Streptocycline (100 ppm) solution for 1hour.
5. Collect and destroy the infected plants regularly.
6. Apply one spray of Streptocycline (100 ppm) with the initiation of the disease followed by copper oxychloride (0.3%) or Bordeaux mixture (4:4:50) and repeat at 7- 10 days interval.

c) Viral diseases

10. Tomato mosaic (tomato mosaic virus)

Symptoms

Infected tomatoes develop light and dark green leaf mottling and sometimes distortion of young leaves also occur (Plate 10). The lamina of leaves is much reduced (fern-leaf) and yield is greatly reduced.

Virus

The disease is caused by tomato mosaic virus which belongs to genus *Tobamovirus* and family *Unassigned*. It is an RNA virus with straight rod particles about 300 × 18 nm.

Host range

Although tomato is the most important host of tomato mosaic virus, the virus also occurs naturally on bell pepper and chillies.

Transmission

The virus can persist in soil for many months in infected plant debris from which it can only be eradicated with great difficulty. The virus is transmissible through minor wounds caused by natural mechanical damage to roots and no vector is involved in its transmission. Virus also survives in infected seed (on seed coat).

Management

1. Treatment of freshly harvested seed with dilute hydrochloric acid will inactivate virus contaminating seed coats.
2. Use of virus-free seed and application of strict hygiene can often reduce and sometimes prevent infection.

11. Tomato spotted wilt (tomato spotted wilt virus)

Symptoms

The symptoms of the disease on leaves of affected plants appear as bronzing, curling, necrotic streaks and spots. On leaf petioles, stems and growing tips of infected plants, dark brown streaks also appear. Infected plants remain stunted. Fully ripened fruits show paler red or yellow areas on the skin (Plate 11).

Sometimes mortality of infected plants takes place due to severe necrosis.

Virus

The virus responsible for this disease is Tomato spotted wilt virus which belongs to genus *Tospovirus* and family *Bunyaviridae.*

Host range

The virus has very wide host range including bell pepper, lettuce, pea, tobacco, potato, and a large number of ornamental plant species.

Transmission

The virus is transmitted and spread in nature by thrips. The virus is acquired only by intensive feeding of vector larvae and transmitted by adults.

Management

1. Raise seedlings under protected (fine-mesh netting) conditions to exclude the entry of vector.
2. Nursery beds should be isolated from ornamental plants and other susceptible crops.
3. Keep the surrounding areas of beds free from weeds.
4. Rouge out the infected plants and destroy them immediately.
5. Spray the seedlings in nursery bed with suitable insecticide like Cyantraniliprole 10.26% OD (0.02%) against thrips.

12. Tomato yellow leaf curl (tomato yellow leaf curl virus)

Symptoms

The virus causes severe stunting of young leaves and shoots, resulting in bushy growth of infected seedlings. The leaf margins of the infected plants show upward and inward rolling, interveinal yellowing of leaflets and marked stunting. The virus reduces fruit set considerably, especially when infection takes place at pre-flowering stage.

Virus

The disease is caused by tomato yellow leaf curl Virus. The

virus belongs to genus *Begomovirus* and family *Geminiviridae.* The virus particle size varies from 15-20 x 25-30 nm.

Host range

Tomato is the primary host of the virus. French beans and Petunia are also susceptible to this virus.

Transmission

Whitefly (*Bemisia tabaci*) is the vector of this virus. Non-host plants such as brinjal and datura harbour the whitefly vector in the absence of the main host.

Management

1. In all regions where the virus is endemic, seedlings need to be grown in a glasshouse or a screen house in order to prevent early infection by whitefly feeding.
2. Plant TYLCV non-host trap plants such as squash and cucumber as inter rows to divert whiteflies from tomato.
3. Apply regular sprays of systemic insecticides like imidacloprid 17.8SL (0.0089%), acetamiprid 20%SP (0.01%), neem formulations (2-3 ml/lt), spiromesifen 22.9% SC (0.02%), thiamethoxam 25%WG (0.01%) and fenpropathrin 30%EC (0.0075%) in nursery beds at seedling stage and also under field conditions to check the population of the vector.

Since these all insecticides are systemic in nature, to avoid resistance development use them in rotation.

d) Nematode diseases

13. Root Knot (*Meloidogyne* spp.)

Symptoms

Knot like galls are formed on the roots, which leads to malfunctioning of root system (Plate 12b). On the aerial parts, the leaves turn yellow and growth becomes stunted (Plate 12a).

Predisposing factors

Meloidogyne spp. overwinter in soil in infected plant debris as eggs and juveniles but very wide host range of this nematode ensures its survival from one season to another. Optimum temperature for embryogenesis is 25–30°C. Nematodes remain active in soil with a moisture level of 40–60 per cent.

Pathogen

The disease is caused by *Meloidogyne* spp. Root knot nematodes are secondary, endophytic and gall producing nematodes. The infective stage is the second stage juveniles, which have lightly sclerotized cephalic framework. Adult females are swollen, pyriform, saccate with short neck, vulva is terminal, perineal cuticular pattern of striae is present. Reproduction is parthogenic.

Management

1. Follow crop rotation with cereals and millets especially rice and avoid solanaceous vegetables like chillies, bell pepper, potato, brinjal etc. in rotation.
2. Change the nursery site every year and before sowing treat the nursery bed soil with Formalin (5%).
3. Select only healthy seedlings for transplanting.
4. Grow nematode resistant variety/ hybrid in infested fields for 2–3 years.
5. Grow marigold, castor or Chrysanthemum as intercrop.
6. Three days after transplanting, drench the plants with Velamprime (fluopyram, 500ml/ha) or apply Nimitz (Fluensulfone 2% GR, 1g/ plant) in infested fields.

e) Deficiency diseases

14. Blossom End Rot (Calcium deficiency)

Symptoms

Lesions appear at blossom ends of the fruits. The affected portions of the fruit become sunken, leathery and dark coloured (Plate 13).

Cause

Calcium deficiency.

Predisposing factors

Water imbalance in soil makes the calcium unavailable to the plants. The occurrence of this disorder is more in those fields which have been supplied with higher doses of Nitrogen.

Management

1. At the time of fruit development stage, apply single spray of Calcium carbonate (0.5%).

(B) INSECT-PESTS

1. Cutworm (*Agrotis segetum*) (Lepidoptera: Noctuidae)

Identification

There is a sexual dimorphism in adult moth. The forewings of male moth are brownish pale in colour while that of female these are brown. On each forewing, a semicircular kidney shaped spot is found. The hindwings are white in both the sexes except brown margins in case of the female. The antenna in the case of male is bipectinate while in the female it is setaceous. The larva is stout, hairless, greasy and dark grey in colour.

Damage

Soon after transplanting, cutworms attack the tomato seedlings. The larvae remain hidden during daytime in the soil. During night, they come out and cut the seedling at ground level. The cut branches are sometimes dragged into ground where the leaves are eaten at leisure. The larvae are very voracious eaters but often fell more seedlings than they can consume.

Management

1. Flooding of fields to kill the larvae and pupae.
2. Deep ploughing and tilling of soil to expose larvae and pupuae to be eaten by birds and other natural enemies.
3. Since it is is polyphagous pest so cut and remove weeds adjoining the planted field.
4. Use of completely rotten Farm Yard Manure.

5. In the field where cutworms are known to cause damage in the previous years, drenching with chlorpyriphos 20%EC (0.04%) should be done near the basin of plants.

2. Fruit borer (*Helicoverpa armigera*) (Lepidoptera: Noctuidae)

Identification

The moth is stoutly built (Plate 1a). The colour of forewings in newly emerged male is slightly greenish while in female it is brown which varies from light to dark. With age, the colour of wings becomes grayish. On the apical margin of forewings, wavy lines in the form of light black band are visible and a black spot appears in the centre on upper side of the wing. Seven or eight spots on the margin of forewings while the hindwings are light pale in colour.

Damage

Larva causes the damage which immediately after hatching feeds on foliage, buds and later on flowers. After that they enter small green fruits and small green fruits of 2 cm diameter are preferred (Plate 1b). A single larva damages many fruits. Damage on tomato is severe during April-June.

Management

1. Deep summer ploughing to expose the pupae to the sunlight and natural enemies.
2. Setting up of light traps (12 light traps/ha) during night helps in catching the adults of this pest.
3. Use of sex pheromone traps helps in early detection and prompt action against this pest. For mass trapping of the adults, 50 pheromone traps/ha at 50 metres distance should be installed.
4. Releasing *Trichogramma* spp. @ 1,50,000/ha in three releases at 10-15 days intervals.
5. Use of African marigold as trap crop.
6. Spray application of following insecticides:
    i. Bt formulation @ 0.5kg/ha.
    ii. Spray of HaNPV @ 250 LE/ha at weekly intervals.

iii. Neem formulation @ 2-3 ml/lt of water.

iv. Indoxacarb 14.5% SC (0.02%) or lambda cyhalothrin 5% EC (0.005%) or carbaryl 50%WP (0.1%) or chlorantraniliprole 18.5%SC (0.005%) or flubendiamide 20%WG (0.025%) or novaluron 10%EC (0.02%).

3. Serpentine leafminer (*Liriomyza trifolii*) (Diptera: Agromyzidae)

Identification

Larvae are legless with light yellow while pupae are oval shaped with yellow in colour (Plate 2a). The adult are two winged, very small having dark yellow spot on the thorax and plum-red eyes.

Damage

Larvae cause the damage which feed on leaves in the palisade mesophyll tissues in between a thin papery upper layer (Plate 2b). The pest attacks the tomato crop during April-May and causes severe damage to leaves which ultimately affects the photosynthetic activity of plant.

Management

1. Judicious use of nitrogenous fertilizer reduces the build up of the population of this pest in endemic areas.
2. Severely infested leaves should be removed and destroyed.
3. Application of NSKE (Neem Seed Kernel Extract) @ 4.0 per cent.
4. Natural enemies especially larval and pupal parasitoids are active during July-August to October.
5. Spray application of following insecticides:

   Cytraniliprole 10.26%OD (0.02%) or chlorantrailiprole 8.8% + thiamethoxam 17.5% SC (drenching @50-100 ml/plant).

4. Greenhouse whitefly (*Trialeurodes vaporariorum*) (Hemiptera: Aleyrodidae)

Identification

Whiteflies are small, inconspicuous insect with white coloured wings which remain hidden under leaves. The adults are very small flies while the nymphs are light yellow. Pupal case is very light yellow and surrounded by a waxy fringe. The eggs are attached with a short stalk.

Damage

Damage is caused by both adults and nymphs which insert their mouth parts into the phloem and suck the plant sap from leaves. Due to continuous sucking of sap, the leaves turn yellow and ultimately dry up. When the population is high, the 'honeydew' secreted by this pest leads to the growth of sooty moulds, thus the photosynthetic activity of plant is interfered which ultimately reduces the yield.

Management

1. **Monitoring and mass trapping:** Use yellow sticky traps at the rate of 1 for 20m^2 for mass trapping and at the rate of 1 for 100m^2 for monitoring (Plate 3).
2. **Biological control:** Several biological agents feed on nymphs. Besides some pathogens are also effective against the nymphs of this pest. These include:

   Predators: lacewing larvae, syrphids, coccinellids.

   Parasitoids: *Eretmocerus* sp, *Encarsia* sp.

   Pathogens: *Beauveria bassiana, Verticillum lecani.*
3. Horticultural oils are commonly used against whiteflies.
4. **Cultural Control:** Maintain good sanitation in areas of winter and spring host crops and weeds by destroying and removing all crop residues as soon as possible.
5. **Chemical control:** The pest can be controlled by need based spraying of crop with any of the following insecticides:

   Imidacloprid 17.8%SL (0.01%) or thiomethoxam 25%WG (0.012%) or spiromesifen 22.9%SC (0.025%) or neem based formulations @ 3 ml/litre of water.

5. Mite (*Tetranychus* sp.) (Acari: Tetranychidae)

Identification

The adult mite is very small (0.50 mm long and 0.25 mm in breadth) and red in colour (Plate 4). Eggs are round with very light yellow in colour.

Damage

Damage is caused by nymphs and adults in the dry month of May-June by feeding on underside of leaves. Due to extraction of chlorophyll, leaves show characteristic blotching and bronzing. Gradually the infested leaves dry up.

Management

1. Remove the old and infested leaves and burn them.
2. Try to avoid dry conditions and spray frequently with plain water at least twice a week with sprinkler.
3. Observe the plants regularly for mite population and if incidence is noticed spray the crop with miticide like fenazaquin 10%EC (0.0025%) or spiromesifen 22.9%SC (0.025%).
4. Use of predatory mite, *Phytoseiulus* spp.

6. Fruit fly (*Bactrocera tau*) (Diptera: Tephritidae)

Identification

The adults are light brown in colour with lemon yellow curved vertical markings across the thorax. On the apical margin of the wing grayish brown patches are present. The larvae are pale or reddish white in colour, which tapers anteriorly. The pupa is barrel shaped with dull to deep reddish yellow in colour.

Damage

The female adult inserts her ovipositor and lays eggs inside fruits. Larvae on hatching cause damage by feeding the fruit pulp inside the fruit. Due to their feeding, the fruits start rotting thus render them unsuitable for human consumption.

Management

1. Collection and destruction of the fallen infested fruits regularly on campaign bases.
2. Frequent raking of the soil to expose pupae.
3. Spray application of poison baits as per the details given below:

Any contact insecticide	=	10 ml
Gur/molasses	=	50 gram
Water	=	5 litres

4. Mass trapping of adults using cue lures.
5. Larval parasitoid, *Biosteres dacusii* also attacks the pest in nature.
6. Male sterilization technique.
7. Bagging of fruits in kitchen garden or smaller areas.

7. Pin worm (*Tuta absoluta*) (Lepidoptera: Gelechiidae)

Identification

Adults are grey brown in colour of 5.5 mm size with wing expanse of about 9.5 mm. The body is covered with alternating rings of light and brown scales which are darker in males than females. These have filiform antennae. Larvae are yellow-green in colour with pinkish tinge on the dorsal surface.

Damage

The pest was initially observed in Pune on tomato plants grown in polyhouse and fields during October 2014 (https: //icar.org.in/node/6084). Later on it was reported from other regions of country from South India to North India. The damage is caused by larvae which feed on leaves, stems and green fruits. Larvae also make irregular mines on leaf surface leaving inside the faecal matter. In severe infestation, leaves die off. Attacked green fruits show the puncture marks of pin hole size. The damage by this pest may cause even upto 100% damage.

Management

1. Installation of sex pheromone traps @ 15-25 traps/ha for monitoring and 30-40 traps/ha for mass trapping.

2. Since pupation takes place in soil, so by encouraging clean cultivation, destruction of crop residues and infested fruits, further check in multiplication of this pest takes place.
3. Crop rotation should be followed.
4. Use of bio pesticides:

   Botanicals: Neem based formulations @ 2-3 ml/ litre of water. Neem seed kernel extract @ 4-5%

   Parasitoids: *Trichogramma* spp.

   Predators: Spiders, mired bugs, wasps, lacewings, ants etc.

   Pathogens: Bt, NPV and fungus (*Beauveria bassiana*).

Disclaimer

Though utmost care was taken in recommending only approved Pesticides for management of different diseases and insect-pests, pesticide recommended wrongly if any, may be purely unintended, inadvertent and by oversight. Further no pesticide can be used in commercial scale without CIBRC approved label claim on the target crop.

4

DISEASES AND INSECT PESTS OF BELL PEPPER AND CHILI

Chili and bell pepper (*Capsicum annum* L.) popularly known as "*Lal Mirch*" and "*Shimla Mirch*", respectively, are important solanaceous crops which are grown throughout the world. In India, these crops are grown in most of the states of the country. Diseases and insect- pests are the major limiting factors in the successful cultivation of these crops. A detailed account of diseases and insect-pests affecting these crops are described herein.

(A) DISEASES

a) Fungal diseases

1. Leaf blight and fruit rot (*Phytophthora capsici, P. nicotianae* var. *nicotianae*)

Symptoms

The symptoms of this disease appear as water soaked bleached spots on the leaf lamina (Plate 1a) causing premature defoliation. Similar spots also appear on the fruits and the flesh below the skin become soft (Plate 1b). There is a demarcation between the diseased tissue and healthy portion. In humid conditions, whitish grey mould appears on the rotten fruits. Severely infected fruits sometimes drop down on the field floor.

Pathogen

Two species of *Phytopthora* namely *P. capsici* and *P. nicotianae* var. *nicotianae* have been found associated with the disease. The pathogen(s) belongs to kingdom Chromista, phylum Oomycota, class Oomycetes, order Peronosporales and family Pythiaceae. In *P. capsici* mycelium is hyaline, branched and non-septate but few septa are found in case of old hyphae. The sporangia are hyaline, ovoid to pyriform or sometimes round to lemon shaped, non-pedicillate with a predominant, hemispherical papilla at the apex. The oospores are circular to spherical which are found to germinate either by germ tubes or by stalked or sessile germ sporangia. In *P nicotianae* var. *nicotianae*, the mycelium is hyaline branched and non-septate with branches at right angles. The sporangia are hyaline, non-pedicellate, globose to subglobose with a prominent beak like papilla at the apex. The sporangia germinate by production of zoospores. The oospores are thick walled golden brown in colour.

Predisposing factors

The pathogen is both seed and soil borne. Rain splashed zoospores cause infection of both leaves and fruits. The disease is favoured by moderate temperature ranging from 22.5- 25°C along with high humidity (80%).

Management

1. Follow crop rotation.
2. Use healthy seed and treat the seed with captan (0.3%)
3. Collect fallen infected leaves and fruits regularly and destroy them.
4. Maintain proper drainage of the field.
5. Apply pine needle/ grass mulch on the field floor before the onset of monsoon rains to avoid splashing of the inoculum.
6. Spray the crop with the onset of monsoon rains with sprays of either mancozeb (0.25%) or zineb (0.25%) or copper oxychloride (0.3%) or Bordeaux mixture (4:4:50) at 7-10 days intervals.

2. Cercospora leaf spot (*Cercospora capsici*)

Symptoms

Circular spots with brown borders and grayish white centre appear on the leaf lamina (Plate 2). In favourable weather conditions many such spots appear on the leaf and coalesce. Such severely infected leaves turn yellow and drop down prematurely.

Pathogen

The fungus responsible for this disease is *Cercospora capsici.* The pathogen belongs to kingdom Fungi, phylum Ascomycota, class Deutromycetes, order Moniliales and family Dematiaceae. The fungus is characterized by dark brown stromata, from which arise fasciculate, straight or slightly curved, continuous, simple yellowish brown conidiophores, producing polymorphous, cylindrical, fusoid 1 to 3 septate, acrogeneous yellow to brown conidia broad and tapering towards the apex.

Predisposing factors

The fungus overwinters in the diseased refuse and possibly carried in seed also. Moderate temperature (20–25° C) and high humidity favour disease development.

Management

1. Collect fallen infected leaves and destroy them.
2. Follow crop rotation and ploughing under all infected plant material.
3. Maintain proper drainage of the field.
4. Use healthy seed.
5. Spray the crop with mancozeb (0.25%) or copper oxychloride (0.3%) or zineb (0.25%) or hexaconazole (0.05%) and repeat at 10–14 days interval.

3. Powdery mildew (*Leveillula taurica* (Lev) Arnaud (anamorph = *Oidiopsis sicula* Scalia Syn. *Oidiopsis taurica* (Lev) Arnaud)

The disease appears in both open as well as protected conditions but it is more severe under protected cultivation conditions.

Symptoms

White to greycoloured spots appear on the under surface of the leaves (Plate 3) while the upper surface of such spots show yellow coloured spots with brown centres. Severely infected leaves curl upward and drop down prematurely due to senescence of the leaves.

Pathogen

The disease is caused by *Leveillula taurica* (Lev) Arnaud (anamorph = *Oidiopsis sicula* Scalia Syn. *Oidiopsis taurica* (Lev) Arnaud). The pathogen belongs to kingdom Fungi, phylum Ascomycota, class Ascomycetes, order Erysiphales and family Erysiphaceae. Its mycelium is endophytic and permeates the host tissues and often covers the whole plant in a white or pale buff membraneous or crustaceous stroma. Perithecia form within the stroma but very rarely. Appendages hyphae like, short, indistinctly branched and colourless to olivaceous brown. Usually there are 20-35 asci/ perithecium and each perithecium contains two ascospores.

Symptoms

White to grey-coloured spots appear on the under surface of the leaves (Plate 3) while the upper surface of such spots show yellow coloured spots with brown centres. Severely infected leaves curl upward and drop down prematurely due to senescence of the leaves.

Predisposing factors

The fungus is obligate parasite in nature. It has wide host range (cultivated and wild), which ensures the survival of the pathogen from one season to another. Higher relative humidity during night in comparison to day and 30° C temperature favours disease development and its progress.

Management

1. Collect and destruct the fallen infected leaves regularly.
2. Mycelium being internal, symptoms appear on lower surface of the leaf lamina, the disease can be kept in check only by application of systemic fungicides like teubuconazole (0.05%) or hexaconazole (0.05%) or flusilazole (0.04%) and repeat at 10-14 days intervals.

4. Die-back and Anthracnose (*Colletotrichum capsici, C. gloeosporoides*)

Symptoms

C. capsici: Branches show die-back symptoms and become straw coloured. Mature green fruits and turning red are more susceptible to the pathogen. Fruits have dark sunken spots along with pink areas of fungal fructifications (Plate 4).

C. gloeosporoides: Small circular or elongated slightly sunken bleached spots appear on the fruits which bear light black, brown or orange coloured acervuli either in the rings or in a scattered fashion.

Pathogen(s)

The disease is caused by two fungi namely *Colletotrichum capsici* and *C. gloeosporoides*. The pathogen(s) belong to kingdom Fungi, phylum Deutromycota, class Coelomycetes, order Melanconiales and family Melanconiaceae.

In *C. capsici* (Tel: *Glomerella cingulata*) the mycelium is septate. Acervuli are round and elongated in shape. Setae are scattered, brown, 1-5 septate. Conidiophores are short, hyaline septate or aseptate. Falcate conidia are one celled with acute apices and narrow truncate base.

In *C. gloeosporioides*, septate mycelium formed brown to orange acervuli. Acervuli bearing setae are rarely observed in the culture; however, on fruits the acervuli are subepidermal with light to dark brown setae disrupting outer epidermal cell walls of the host. Conidiophores are hyaline bearing oblong to cylindrical conidia with obtuse ends having 1-3 oil globules.

Predisposing factors

The pathogen overwinters both in infected seed and plant debris. Disease development is favoured by free moisture on the surface of fruits in the form of dew or 100 per cent relative humidity along with 28° C temperature.

Management

1. Collect and burn the infected plant debris.
2. Obtain seed from healthy fruits.
3. Treat the seed with captan (0.3%).
4. Spray the crop with mancozeb (0.25%) or zineb (0.25%) or copper oxychloride (0.3%) or difenconazole (0.05%) or hexaconazole (0.05%) or tebuconazole (0.05%) and repeat at 10-14 days interval.

b) Bacterial diseases

5. Bacterial spot (*Xanthomonas campestris* pv. *vesicatoria*)

Symptoms

The symptoms of the disease appear on leaves as water-soaked spots which are either circular or irregular in shape. Later these spots become necrotic with brown centre are surrounded by thin chlorotic borders (Plate 5a). Upper surface of these spots is generally sunken while slightly raised on the under surface. Infected leaves turn yellow and drop down prematurely. On fruits, raised spots appear which often crack in the centre (Plate 5b).

Pathogen

The disease is caused by bacterium *Xanthomonas campestris* pv. *vesicatoria*. The bacterium is described in the chapter of tomato diseases.

Predisposing factors

The pathogen is both seed and soil borne. Disease development is favoured by moderate temperature and high relative humidity. Frequent rains help in the progress of the disease.

Management

1. Collect fallen infected leaves and destroy them.
2. Follow crop rotation and avoid tomato in rotation.
3. Use healthy seed.
4. Spray the crop with copper oxychloride (0.3%) or Bordeaux mixture (4:4:50) and repeat at 7-10 days interval.

6. Bacterial wilt (*Ralstonia solanacearum*)

Symptoms

Infected plants show sudden wilting of foliage, which is succeeded, by a sudden and permanent wilting of entire plants, with a slight or no leaf yellowing.

Pathogen

The disease is caused by *Ralstonia solanacearum*. The bacterium is described in tomato diseases.

Predisposing factors

The pathogen overwinters in soil and it is having a wide host range.Severe wilting is seen at high temperatures of 28° C or higher and high soil moisture.

Management

1. Follow at least three years crop rotation with non-solanaceous crops.
2. Solarize the field soil before planting for at least 45 days.
3. Apply bleaching powder (15 kg/ ha) before transplanting.
4. Use resistant cultivars/ hybrids.

c) Viral diseases

7. Chilli Veinal mottle (chilli veinal mottle virus)

Symptoms

Leaves of infected plants show dark-green mottles, with dark-green spotting, patching and banding adjacent to the veins (Plate 6). Mottling is most conspicuous on young leaves. Affected leaves are generally smaller in size and show varying degrees of distortion. Plants infected early are usually stunted, show dark-green streaks on the stems and branches, and most of their flowers drop before bearing fruits.

Virus

The disease is caused by Chilli veinal mottle virus which belongs to genus *Potyvirus* and family *Potyviridae.* Virus particles are flexuous, rod shaped ranging from 680-900 nm in length.

Host range

Bell pepper, chili and tobacco are the natural hosts of the virus. Some weeds growing around bell pepper fields have been identified as alternate hosts of this virus.

Transmission

The virus is transmitted by different methods like mechanical inoculation, grafting, and aphid (*Aphis craccivora,* and *Myzus persicae*) vectors in non-persistent manner, but not through seed.

Management

1. Inter-cropping of maize with *Capsicum* species and planting maize as a barrier crop around *Capsicum* plantation reduce vector populations and virus incidence.
2. Sprays of mineral oil alone or in combination with reflective mulches like aluminium-painted or silver-coated polyethylene sheets help to delay virus infection.
3. Application of systemic insecticide sprays also help in keeping the vector population in check.

8. Pepper veinal mottle (pepper veinal mottle virus)

Symptoms

Most isolates or strains of the virus induce chlorotic vein banding and leaf mottling, leaf distortion, reduced fruit set and plant stunting (Plate 7). Severe strains of thevirus also induce leaf and stem necrosis.

Virus

The disease is caused by pepper veinal mottle virus. The virus belongs to genus *Potyvirus* and family *Potyviridae*. Virus particles are flexuous, rod shaped ranging from 680-900 nm in length.

Host range

Besides bell pepper and chillies, the virus also infects tomato and brinjal.

Transmission

Different Aphid species like *Aphis craccivora, A. fabae, A. gossypii* and, *Myzus persicae* transmit this disease in non-persistent manner. Wingless stage of the aphids is largely responsible for the spread of this virus. The virus is not transmitted through seed.

Management

1. Removal of virus infected weeds from in and around the nursery bed.
2. Management of aphid vectors with spray application of insecticides like cabosulfan 25%EC (0.025%) or fipronil 5%SC (0.0075%) or imidacloprid 17.8%SL (0.01%) help in

minimizing the establishment of early virus infection in the field.

3. Use immune or resistant cultivars, if available.

9. Pepper mild mottle (pepper mild mottle virus)

Symptoms

Mild leaf mottling, malformed mottled fruits having sunken necrotic spots and stunting of the plants are the characteristic symptoms of this disease. Severity of the disease is more on those plants which are infected by virus early.

Virus

The disease is caused by pepper mild mottle virus. It belongs to genus *Tobamovirus* and family *unassigned*. Virus particles are rigid, rod shaped and approximately 300 nm in length.

Host range

Besides bell pepper and chillies, the virus also infects tomato and brinjal.

Transmission

The primary source of infection of this virus is infected seed as well as plant debris of the previous crop. During transplanting, seedlings can get infection by mechanical contamination from the contaminated seed coats as well as through different cultural operations.

Management

1. Identify infected plants as early as possible and destroy them. This practice is essential to keep the secondary spread of the disease in check.
2. Use healthy seed collected from disease free plants.
3. Use such cultural practices in which there is minimum contact in between plants, workers and equipments.

d) Nematode diseases

10. Root Knot (*Meloidogyne* spp.)

Symptoms

Knot like galls are formed on the roots, which leads to malfunctioning of root system. On the aerial parts, the leaves turn yellow and growth becomes stunted.

Pathogen

The disease is caused by *Meloidogyne* spp. The pathogen is described in tomato chapter.

Predisposing factors

Meloidogyne spp. overwinter in soil in plant debris as eggs and juveniles but wide host range ensures its survival from one season to another. Optimum temperature for embryogenesis is 25-30° C. Nematodes remain active in soil with a moisture level of 40-60 %.

Management

1. Follow crop rotation with cereals and millets especially rice and avoid solanaceous vegetables like tomato, potato, brinjal etc. in rotation.
2. Change the nursery site every year and treat the nursery bed soil before sowing with Formalin (5%).
3. Select gall free seedlings before transplanting to check the spread of the nematodes to uninfected fields.
4. Grow nematode resistant variety/ hybrid in infested fields for 2-3 years.
5. Grow marigold, castor or Chrysanthemum as intercrop.

(B) INSECT-PESTS

1. Green peach aphid (*Myzus persicae*) (Hemiptera: Aphididae)

Identification

The adults are greenish, smooth looking aphids whereas nymphs are pale yellowish green in colour with three dark lines on the back of abdomen (Plate 1).

Damage

Both nymphs and adults cause damage by inserting their mouth parts in the phloem of leaves. Due to sucking of cell sap, the plants start drying and the growth is reduced thus plant remains stunted. Formation of flowers and fruits is adversely affected. This aphid acts as vector of many plant viruses.

Management

1. Optimum use of nitrogenous fertilizers.
2. Since this aphid also develops on non-crop hosts so remove the weeds and crop residues before planting of crop.
3. Many biological control agents (syrphids, coccinellids, chrysopids) help in reducing the population of this aphid.
4. Use of neem based formulations @ 2-3 ml/litre of water.
5. With the initiation of aphid attack, spray the crop with cabosulfan 25%EC (0.025%) or fipronil 5%SC (0.0075%) or imidacloprid 17.8%SL (0.01%) and repeat as per requirement.

2. Tobacco caterpillar (*Spodoptera litura*) (Lepidoptera: Noctuidae)

Identification

The adults are 2.5 cm in length with velvety black-white wavy lines on forewings. The larvae are yellowish green with lateral white bands.

Damage

It feeds on tobacco, castor, tomato and various crops. The damage is done by larva, which feeds on leaves and tender shoots.

Management

1. The eggs masses and larvae in gregarious phase can be destroyed mechanically.
2. Mechanical removal of larvae.
3. Deep summer ploughing to expose the pupae to the sunlight and natural enemies.
4. Setting up of light traps (1-2 light traps/ha) during night helps in catching the adults of this pest.

5. Use of sex pheromone traps helps in early detection and prompt action against this pest. For mass trapping of the adults, 50 pheromone traps/ha should be installed.
6. Growing of caster as trap crop.
7. Spray application of any one of the following insecticides:
   - Bt formulation @ 0.5 kg/ha.
   - Spray of HaNPV @ 250 LE/ha at weekly intervals.
   - Neem formulation @ 2-3 ml/lt of water.
   - Cytraniliprole 10.26OD (0.02%).

3. Termite (*Odontotermes obesus*) (Isoptera: Termitidae)

Identification

These insects can be spotted out by presence of mud layer around or near the plant. These are soft bodied, light yellow coloured insects. They have caste system within colony. The queen is the largest in size. The soldier's have strong jaws, while workers are smaller.

Damage

This pest is found throughout the year but the incidence during monsoon is low. Damage is done by workers, which gnaw the roots and then move upwards making tunnels on m ain stem. The attacked plant withers, dries up and ultimately dies. They eat up live as well as dead plant materials. Drying, wilting or withering and falling of plants normally occur especially when termites attack the main root system.

Management

1. Avoid application of partially rottened green manure and farmyard manure.
2. **Cultural practices:** The colony may be destroyed by excavating the mound to remove the queen but this may not work for groups that have the ability to quickly replace the lost queen. Excavating the mound may be an extremely laborious task, since the queen may be concealed at some depth. Other cultural practices include the following:

- Deep ploughing to destroy the termites' nests, runways, and tunnels and to expose them to predators, such as ants, birds, chickens, etc.
- Heavy irrigation helps in drive away the pest.
- Remove plant residues and other debris especially moist and decaying wood.

3. **Biological control**

1. Termites have a host of natural enemies that prey upon the colony-establishing alates (winged termites) during swarming and as pairs dig-in to establish new colonies, or on foraging termites or the individuals in the nest itself.
2. Enhancing ant predation by putting out baits for predatory ants (e.g. fish bones) within the crop or transplanting of ant nests into the crop can interfere with termite foraging.
3. The use of entomopathogenic (insect attacking) fungi such as *Beauveria bassiana* and *Metarhizium anisopliae* are reportedly effective against adult termites.

4. **Chemical control**

Chlorpyriphos @ 3-5 ml/lt of water.

Disclaimer

Though utmost care was taken in recommending only approved Pesticides for management of different diseases and insect-pests, pesticide recommended wrongly if any, may be purely unintended, inadvertent and by oversight. Further no pesticide can be used in commercial scale without CIBRC approved label claim on the target crop.

5

DISEASES AND INSECT- PESTS OF BRINJAL

Brinjal (*Solanum melongena* L.), also referred to as eggplant, aubergine or guinea squash, is among the important vegetables which is grown almost worldwide. The raw vegetable contains only 15 calories/ 100 g but its calorie value rises sharply when it is fried. During cultivation it is affected by several diseases and insect-pests, which not only reduce the yield but also impair the fruit quality. In this chapter, a detailed account of various diseases and insect-pests affecting this crop are described along with their management practices.

(A) DISEASES

a) Fungal diseases

1. Phomopsis Blight and Fruit Rot (*Phomopsis vexans*)

Symptoms

The disease affects the crop in all of its growth stages. In nursery beds, seedlings are damped off at soil line. In field, circular grey spots with light colored centre appear on the leaves (Plate 1a). When many such spots appear on the leaves, they coalesce and infected leaves turn yellow and fall down pre maturely. On stems, the symptoms of the disease appear as elongated, blackish-brown lesions. On fruits, pale sunken spots with black pycnidia develop which may cover the entire fruit surface under favourable weather conditions (Plate 1b).

Pathogen

The disease is caused by *Phomopsis vexans*. (Tel: *Diaporthe vexans*). The pathogen belongs to kingdom Fungi, phylum Ascomycota, class Deutromycetes, order Sphaeropsidales and family Sphaeropsidaceae. The mycelium is hyaline and septate. The conidiophores (phialides) in the pycnidium are hyaline, branched and some times septate. Two types of conidia viz., alpha and beta are produced. At 10–16° C, beta conidia and at 25–28° C, the alpha conidia are produced. These two forms of conidia get inter-converted when subjected to specific temperature. Alpha conidia are hyaline, one celled, and sub cylindrical while Beta conidia are stylospores is filiform, curved, hyaline and septate. These spores normally do not germinate. But inoculation of host with beta conidia caused interveinal necrosis. Perithecia are beaked, carbonaceous, sinuate, and with irregular ostiole. Asci are clavate, sessile and contain eight ascospores which are hyaline, narrowly ellipsoid to bluntly fusoid, one- septate and constricted at the septum.

Predisposing factors

The pathogen is both seed and soil borne. Warm (25°C) temperature and wet weather conditions favour infection and subsequent disease development.

Management

1. Since the pathogen overwinters in infected plant debris, collect and burn the infected plant debris.
2. Follow atleast 2 years crop rotation.
3. Use healthy seed and treat the seed with carbendazim (0.2%).
4. Spray the crop with carbendazim (0.1%) or zineb (0.25%). Repeat sprays at 10 days interval.

2. Sclerotinia wilt (*Sclerotinia sclerotiorum*)

Symptoms

The symptoms of the disease appear as circular to elongate water-soaked lesions closer to the inflorescence on the branches followed by watery soft rot. When the infection occurs at the base of the main stem, wilting of such plants takes place. In cool humid conditions, the mycelium emerges out on infected above ground plant parts and studded with black sclerotia of varying sizes. Infected fruits start rotting and in rotting tissue large number of sclerotia is seen.

Pathogen

The disease is caused by *Sclerotinia sclerotiorum.* The pathogen belongs to kingdom Fungi, phylum Ascomycota, class Discomycetes, order Helotiales and family Sclerotiniaceae. Mycelium is hyaline, cottony, branched, consisting of closely septate hyphae. Microconidia (spermatia) are produced on short lateral branches of the vegetativc mycelium in chains. When food supply is exhausted Sclerotia are formed. These sclerotia are first white in colour and then turn black. The sclerotia also germinate by producing stalked apothecia (1-5/ sclerotium), when proper humidity and light conditions are provided to over wintered sclerotia. The best appothecial development takes place at 7-11° C. Apothecia are brown in colour and are round or lobate type. Asci are cylindrical. Each ascus contains eight ascospores.

Predisposing factors

The pathogen overwinters in soil in the form of sclerotia. Cool (15 to 20° C) and humid weather favours disease development.

Management

1. Destruction of infected plant debris.
2. Follow paddy or maize rotation.
3. Spray the crop with carbendazim (0.1%) or zineb (0.25%). Repeat at 10 to 14 days interval.

3. Phytophthora Fruit Rot (*Phytophthora parasitica*)

Symptoms

The symptoms of the disease appear as rotting of the fruits from the apical portion. In moist weather, infected fruits are covered with fluffy mycelium of the fungus (Plate 2). Such infected fruits drop down on the ground prematurely.

Pathogen

The disease is caused by *Phytophthora parasitica.* The pathogen(s) belongs to kingdom Chromista, phylum Oomycota, class Oomycetes, order Peronosporales and family Pythiaceae. Pathogen is described under tomato diseases.

Predisposing factors

The fungus overwinters in the soil in the form of oospores or chlamydospores. The disease assumes serious proportions when moisture or rainfall is plentiful coupled with temperature ranging from 22.5-25° C during the cropping season.

Management

1. Follow crop rotation.
2. Collect and destroy the infected plant debris.
3. With the initiation of the disease spray the crop with zineb (0.25%) or Bordeaux mixture (4:4:50). Repeat sprays at 7-10 days interval.

4. Alternaria leaf spot (*Alternaria melongenae*)

Symptoms

The symptoms of the disease appear as brown spots which are irregular in shape. These spots enlarge in size by forming concentric rings (Plate 3). In favourable weather conditions many spots may coalesce to form larger necrotic patches. Sunken necrotic spots also develop on fruits.

Pathogen

The disease is caused by *Alternaria melongenae.* The pathogen belongs to kingdom Fungi, phylum Ascomycota, class Deutromycetes, order Moniliales and family Dematiaceae. No sexual stage of the fungus is known. The mycelium consists of septate branched, light brown hyphae which become darker with age. The size of conidia ranges from 31.5–89.5 × 10.10–5.21 µm.

Predisposing factors

The pathogen overwinters in soil as mycelium in the infected plant debris. Optimum temperature for disease development is 28–30° C. Moisture is essential for spore germination.

Management

1. Collect and destroy the infected plant debris.
2. With the initiation of the disease, spray the crop with zineb (0.25%) or Bordeaux mixture (4:4:50) and repeat at 14 days interval.

b) Bacterial diseases

5. Bacterial Wilt (*Ralstonia solanacearum*)

Symptoms

The leaves of the infected plants suddenly droop down without yellowing. With the progress of the disease whole plant wilts.

Pathogen

The disease is caused by *Ralstonia solanacearum*. The bacterium is described in tomato chapter.

Predisposing factors

The bacterium has a wide host range and can also survive in moist soil for longer periods in the absence of plant debris. The disease usually develops when mean temperature is 20° C and more severe wilting is observed at higher temperatures of 30° C or more along with high soil moisture. Disease is further aggravated if soil is infested with root knot nematode.

Management

1. Follow at least three years crop rotation with non-solanaceous crops.
2. During summer months, solarize field soil before planting for at least 45 days.
3. Apply bleaching powder (15 kg/ ha) before transplanting.
4. Use resistant variety/ hybrid if available.
5. At the time of field preparation, apply Thimet/ Phorate (12.5-15 kg/ ha) or Nimitz (25-30 kg/ha immediately after transplanting) or Velamprime (Fluopyram, 500ml/ ha).

c) Phytoplasma diseases

6. Little leaf (*Phytoplasma*)

Symptoms

The disease is characterized by smalling of leaves and shortening of nodes and internodes, which gives the plant a bushy appearance. The fruiting in infected plants is rare.

Pathogen

The disease is caused by genus *Phytoplasma* which belongs to group Mollicutes, family Acholeplasmataceae. *Phytoplasma* bodies (230-770 nm) are found associated with the phloem of roots of the host. Each *Phytoplasma* body contains ribosomes and strands of nuclear material surrounded by 16.5 nm wide triple layered unit membrane. Electron micrographs revealed pleomorphic *Phytoplasma* bodies in the sieve elements of the affected plants.

Predisposing factors

In nature, the *Phytoplasma* perennates on the weed hosts like *Datura* spp.and *Vinica rosea* and several others. The disease is transmitted by grafting and through leafhopper vector, *Hishimonas phycitis.*

Management

1. Destruction of weeds like *Datura* spp. and *Vinica rosea* from in and around the field.
2. Before transplanting, dip seedling in tetracycline (0.05%).
3. Apply Phorate (1kg/ha) in soil to keep the vector population under check.

d) Nematode disease

7. Root Knot (*Meloidogyne* spp.)

Symptoms

Knot like galls are formed on the roots, which leads to malfunctioning of root system. On the aerial parts, the leaves turn yellow and growth becomes stunted.

Pathogen

The disease is caused by *Meloidogyne* spp. The pathogen is described in tomato chapter.

Predisposing factors

Meloidogyne spp. overwinter in soil in infected plant debris as eggs and juveniles but wide host range ensures its survival from one season to another. Optimum temperature for embryogenesis is

25-30° C. Nematodes remain active in soil with a moisture level of 40-60 per cent.

Management

1. Follow crop rotation with cereals and millets especially rice and avoid solanaceous vegetables like chillies, bell pepper, potato, brinjal etc. in rotation.
2. Change the nursery site every year and apply Carbofuran granules (3g/m^2).
3. Select gall free seedlings before transplanting to check the spread of the nematodes to uninfected fields.
4. Grow nematode resistant variety/ hybrid in infested fields for 2-3 years.
5. Grow marigold, castor or Chrysanthemum as intercrop.
6. Apply Thimet/ phorate granules (1.25-1.5 kg/ bigha) before transplanting or Nimitz (25 30 kg/ha immediately after transplanting) or drench Velamprime (Fluopyram, 500ml/ ha) 2-3 days after transplanting.

(B) INSECT-PESTS

1. Brinjal shoot and fruit borer (*Leucinodes orbonalis*) (Lepidoptera: Crambidae)

Identification

The eggs are creamy white. Full grown caterpillars are light pink in colour. Adults are white and medium sized having light brown spots on forewings. Hind wings are opalescent with black dots along the margins. Wing span is 22-26 mm.

Damage

The damage is done by larvae which bore into tender shoots, flower buds and ultimately into fruits. Fruits show circular exit holes. These fruits become unfit for human consumption and lose market value. Infestation up to 70 per cent may be recorded.

Management

1. **Mass Trapping:** Mass trapping of adult male moths using pheromones in water trough traps @ 40-50 traps per hectare.

2. Destruction of affected shoots and fruits is useful in killing larvae and pupae
3. The small and stout brinjals are more susceptible to the pest attack compared to long and slender varieties so long and slender varieties should be encouraged.
4. Same crop should not be planted continuously in the same field year after year.
5. Destruction of infested shoots along with the larvae at 15 days interval.
6. **Biological control:**
   - **Predators:** *Campyloneura* sp *(a bug), Cheilomenes sexmaculata (a ladybird beetle), Coccinella septempunctata (seven spotted ladybird beetle), Brumoides suturalis* (three striped ladybird).
   - **Parasitoids:** Periodic releases of *Trichogramma chilonis* @ 1,50,000 parasitised eggs/ha
   - Entomopathogens Fungus
   - Nuclear polyhedrosis virus
   - Bt formulations @ 500g/ha
7. **Chemical control:** Spray chlorantraniliprole 18.5% SC (0.1%) or chlorpyriphos 20%EC (0.02%) or cypermethrin 10%EC (0.02%) or cypermethrin 25%EC (0.001%) or lambda cyhalothrin 5%EC (0.003%) or quinalphos 2.5%EC (0.01%) or thiacloprid 21.7%SC.

   (0.003%) orthiocarb 75%WP (0.1%)). Besides, the neem based formulations @ 2-3 ml/lt of water.

2. Brinjal hadda beetles (*Henosepilachna vigintioctopunctata, H. duodecastigma, H. demurili*) (Coleoptera: Coccinellidae)

Identification

Adult beetles are about 8 to 9 mm in length and 5 to 6 mm in breadth. The larvae (grubs) are about 6 mm, yellow, with six rows branched spines. The three species of adult beetles can be identified as per the details given below:

H. vigintioctopunctata: The adult beetles are deep copper coloured having 14 black spots on each elytron whose tip is somewhat pointed.

H. duodecastigma: The adults are deep copper coloured with 6 black spots on each elytron whose tip is more rounded.

H. demurili: Adult beetles are dull in appearance and light copper coloured with each elytron bearing 6 black spots surrounded by yellow margins.

Damage

Damage is done by adults and grubs which feed on chlorophyll tissue between veins. The leaves ultimately fall off. In severe infestation, the leaves are completely skeletonized.

Management

1. Collection and destruction of various stages of the pest.
2. Larval parasitoids such as *Pediobius* sp. is active in nature.
3. **Chemical control:** The pest can be controlled by spraying the crop with cypermethrin 25%EC (0.0075%), or combi-product: deltamethrin1% + trizophos 35%EC. Besides these, neem based formulations @ 2-3 ml/lt of water can also be used for its management.

3. Leafhopper: *Amrasca biguttula biguttula*, *A. devastans* (Hemiptera: Cicadellidae)

Identification

The eggs are light yellow, pear shaped. Nymphs are wingless and pale green. Adults are very small (about 2-3 mm in length), wedge shaped and pale green in colour with black round spot on the posterior part of each forewing.

Damage

Damage is caused by both nymphs and adults which suck cell sap from the leaves especially from the lower surface of the leaves. During the process of feeding, they also inject toxic saliva. Due to continuous sucking of cell sap by many nymphs and adults, yellow spots appear which further leads to crinkling, curling and bronzing of leaves. The leaves start drying and the plant shows a burning condition of all the leaves popularly known as 'hopper burn'.

Management

1. Seed treated with imidacloprid (3g/kg) gives protection up to 40–50 days after sowing.
2. Predators, like *Coccinella* spp, *Chrysoperla* sp., spiders etc. predate upon nymphs and adults of this pest.
3. Spray Carbaryl 50%WP (0.1%) or cypermethrin 25%EC (0.0075%) or thiometon 25%EC (0.025%) or combi product: beta cyfluthrin 8.49%+imidacloprid 19.81% OD. Besides, neem based formulations @ 2–3 ml/lt of water can be used. Same insecticides should not be repeated.
4. Neem seed kernel extract (5%) also helps in reducing the population of this insect.

Besides, the above mentioned insects, some other insects like whitefly, red spider mite, stem borer, leaf roller etc also attack this plant.

Disclaimer

Though utmost care was taken in recommending only approved pesticides for management of different diseases and insect-pests, pesticide recommended wrongly if any, may be purely unintended, inadvertent and by oversight. Further no pesticide can be used in commercial scale without CIBRC approved label claim on the target crop.

6

DISEASES AND INSECT-PESTS OF CUCURBITS

Cucurbits are the members of family cucurbitaceae which constitute the largest group of tropical, subtropical vegetables. The family cucurbitaceae contains atleast 9 genera and of them 16 species are cultivated as vegetables. During cultivation several diseases and insect-pests attack these crops, which reduce both quantity as well as quality. In this chapter important diseases and insect pests are described along with their management.

(A) DISEASES

a) Fungal diseases

1. Downy mildew (*Pseudoperonospora cubensis*)

Symptoms

On leaves, pale green to yellow angular spots appear which are limited by leaf veins while there corresponding undersurface is covered by a grey to brownish fungal growth comprising of sporangia and sporangiophores of the pathogen (Plate 1). On the upper leaf surface, lesions turn necrotic brown from the center outwards.

Pathogen

The fungus responsible for this disease is *Pseudoperonospora cubensis*. The pathogen belongs to kingdom Chromista, phylum Oomycota, class Oomycetes, order Peronosporales and family Peronosporaceae. The mycelium is hyaline, coenocytic and intercellular. Sporangiophores develop in groups of 1-5 and are bulbous at base and are dichotomously branched in the upper third of their growth. Sporangia form singly on subacutely shaped sporuliferous tips, are pale grey to purple in colour, ovoid to elliptical in shape, possesses a thin wall with a papilla at the distal end. They germinate by producing zoospores. Oospores are produced rarely and are spherical, rarely obovoid to ellipsoid, light yellow to pale yellow with a smooth wall.

Predisposing factors

The pathogen infects many cucurbitaceous hosts (wild and cultivated). In these plants, pathogen overwinters as mycelium and sporangiophores during the off-season. Oospores are not commonly formed. Moderate temperature (20 ± 2° C) along with high precipitation and relative humidity (80%) helped significantly in disease development.

Management

1. Collect and destroy the infected plant debris and cucurbitaceous weeds from in and around the field.
2. Maintain proper distance among plants to prevent the buildup of favourable microclimate in plant canopy.
3. Spray the crop with cymoxanil + mancozeb (0.25%) or azoxystrobin (0.1%) alone or in combination with chlorothalonil (0.2%) followed by sprays of zineb (0.25%) or Bordeaux mixture 4:4:50 and repeat at 10 days interval.

2. Powdery mildew (*Sphaerotheca fuliginea; Erysiphe cichoracearum*)

Symptoms

The symptoms of the disease appear as white, floury patches on leaves (Plate 2), stems and other succulent parts of the infected plants. In *S. fuliginea,* the spots are dirty brown in colour whereas white in *E. cichoracearum.*

Pathogen(s)

Two fungi namely *Sphaerotheca fuliginea and Erysiphe cichoracearum* are associated with this disease. The pathogen(s) belongs to Kingdom Fungi, phylum Ascomycota, class Ascomycetes, order Erysiphales and family Erysiphaceae. Conidia of *S. fuliginea* are formed in long chains. They often show fibrosin bodies and are ellipsoidal to barrel shaped. Cleistothecia are scattered to densely gregarious. Appendages are myceloid. Each cleistothecium contains single ascus with eight ascospores. In *E. cichoracearum*, the conidia are single celled, hyaline and barrel shaped. Cleistothecia are globose, dark with hyaline to dark brown mycelioid appendages. Each cleistothecium contains 8 to 18 asci and each ascus contains 2–3 ascospores.

Predisposing factors

Both pathogen either overwinter as cleistothecia (wherever cleistothecia are produced) or on perennial cucurbitaceous weeds/ indoors cultivation in the area as conidia. The disease is favoured by moderate to warm temperatures (25° C) along with high relative humidity (95%) in case of *S. fuliginea* while up to 70 per cent in *E. cichoracearum*. Reduced sunshine hours will further aggravate the situation.

Management

1. Collect and burn the infected plant debris.
2. With the initiation of the disease spray the crop with fungicides like carbendazim (0.05%), azoxystrobin (0.1%) or azoxystrobin + chlorothalonil (0.2%) and repeat at 10–14 days interval.

Precaution: Cucurbits are sensitive to sulphur, don't spray sulphur fungicides for the control of this disease.

3. Anthracnose (*Colletotrichum orbiculare*)

Symptoms

On leaves, light brown to reddish brown spots appear which are rough and circular in shape (Plate 3a). Similar spots also appear on fruits which are water-soaked roughly circular and always sunken (Plate 3b), with dark borders. The colour of older spots changes to dark brown to black. In moist weather these spots are covered with pinkish spore mass.

Pathogen

The disease is caused by *Colletotrichum orbiculare*. The pathogen belongs to kingdom Fungi, phylum Deutromycota, class Coelomycetes, order Melanconiales and family Melanconiaceae.

The conidia are one celled, hyaline, oblong to ovate and disseminated by water. The fungus produces black stromata, bearing black setae and hyaline conidiophores on the host surface. Conidia are produced successively by budding at the tips of conidiophores accumulating in pinkish mass. The teleomorph *Glomerella lagenarium* is rarely found in nature.

Predisposing factors

The pathogen is both soil and seed borne. Cucurbitaceous weed hosts also act as the source of perennation. Optimum temperature for spore germination is 22 -27° C and 100 per cent RH or free moisture on the surface of leaves/ fruits. Moist wind or rain splashes disseminate the conidia from one plant to another. Conidia are also carried on the implements from one place to another.

Management

1. Collect and destroy the infected plant debris.
2. Use healthy seed for sowing.
3. Treat the seed with carbendazim (0.2%).
4. Spray the crop with carbendazim/ thiophanate methyl (0.1%) or zineb (0.25%). Repeat sprays at 10-14 days interval.

b) Bacterial diseases

4. Angular leaf spot (*Pseudomonas syringae* pv. *lachrymans*)

Symptoms

On leaves, the symptoms of the disease appear as water soaked spots which are delimited by veins and become angular (Plate 4). Diseased spots and dead tissues are covered with crust of bacterial ooze. Infected tissue may drop down leaving shot holes in the leaves. Small, circular brown to black superficial spots also appear on fruits.

Pathogen

The bacterium associated with this disease is *Pseudomonas syringae* pv. *lachrymans*. The bacterium belongs to kingdom

Prokaryotae, division Gracilicutes, class Proteobacteria and family Pseudomonadaceae. It is Gram negative, aerobic, non-spore forming, straight rod shaped bacterium motile with 1 to 5 polar flagella. It is non-nitrate reducing and secrete protease. Optimum temperature for growth is 25-30° C.

Predisposing factors

The bacterium is both seed and soil borne. Plenty of moisture and moderate temperature (20-24° C) are optimum for disease development and spread.

Management

1. Rotate the field regularly and avoid cucubitaceous crops in rotation.
2. Select healthy seed produced in dry areas.
3. Collect and destroy the infected plant debris.
4. Treat the seed by dipping in Streptocycline (100 ppm) solution for 30 minutes and dry in shade.
5. With the initiation of the disease, spray the crop with copper fungicides like copper oxychloride (0.3%) or Bordeaux mixture (4:4:50) and repeat at 7 to 10 days interval.

c) Viral diseases

5. Cucumber mosaic (cucumber mosaic virus)

Symptoms

Many strains of the virus sometimes cause symptomless systemic infections. The most common symptoms include severe mosaic mottling, chlorosis, necrosis and distortion in leaves (fern-leaf and shoestring symptoms in tomatoes) (Plate 5) and fruits. Occasionally, the infection may result in necrosis and cell death in tomatoes or may cause the amelioration of symptoms.

Virus

The disease is caused by cucumber mosaic virus. The virus belongs to genus *Cucumovirus* and family *Cucmoviridae*.

Host range

The virus has very wide host range and infects cucumber, bell pepper, tomato, cucurbits, okra, celery, sugar beet, mustard, turnip rape, watermelon, squash, pumpkin, carrot, lettuce, beans, radish, brinjal, potato and broad bean.

Transmission

Aphids are the vector of this disease and can transmit the virus in a non-persistent manner. The most common aphid vectors are *Myzus persicae* and *Aphis gossypii.* The virus is stylet-borne and all stages of aphids can act as efficient vectors. Other means of virus transmission are mechanical inoculation and through seed.

Management

Since the virus is transmitted by over 80 aphid species, the diverse behaviour of various aphid vectors can greatly reduce the impact of insecticides. However, some other strategies can minimize the losses caused by the virus.

1. Planting of barrier crops around the field, use of yellow sticky traps, or application of an aphid deterrent material like aluminium foil boards on the ground can help in reducing the disease incidence.
2. The use of virus-free seeds together with the eradication of virus reservoirs can be effective in controlling CMV.
3. In the field, virus spread can also be reduced by the application of neem formulations (2-3 ml/lt).

6. Squash mosaic (squash mosaic virus)

Symptoms

In melon, symptoms on leaves include green vein banding, mosaic, mottle, ring patterns, and leaf deformations and enations. In some varieties, no symptoms appear on fruit, while in others, mottling and loss of netting can be seen on the fruit surface. In squash, mosaic and distortion of the leaves can be seen. Mottled and malformed symptoms may appear on fruits.

Virus

The disease is caused by squash mosaic virus. The virus belongs to the genus *Comovirus* and family *Comoviridae.*

Host range

The principal host of squash mosaic virus is melon, but the virus or its strains may also infect other cultivated cucurbits, such as summer and winter squash, pumpkins and cucumber. Other natural hosts are mainly weeds in the family cucurbitaceae.

Transmission

The virus is mostly found in those areas where cucurbit seeds are produced and seeds play a key role in the transmission of the virus in cultivated cucurbits. The main method of widespread dispersal in most cultivated cucurbits is through the exchange or trade of infected seeds from one country to another. Many Aphid species are the insect vectors.

Management

1. As soon as infected plants are discovered, the infected seedlings should be removed to prevent additional spread.
2. Use of certified virus-free seed is an effective means of limiting the virus.
3. In the field, virus spread can be reduced by the application of neem formulations (2-3 ml/lt).

7. Zucchini yellow mosaic (zucchini yellow mosaic virus)

Symptoms

In cucurbit crops, the virus cause severe symptoms of stunting, yellowing, necrosis (occasionally), mosaic, leaf deformations (blisters, shoe stringing), fruit discolorations and deformations. Sudden wilting of the plants can also be observed due to infection of some strains in melons and in mixed infection with cucumber mosaic virus on cucumbers.

Virus

The disease is caused by zucchini yellow mosaic virus. The virus belongs to the genus *Potyvirus* and family *Potyviridae.*

Host range

The virus causes severe diseases in Zucchini, cucumber, pumpkin and watermelons.

Transmission

Zucchini yellow mosaic virus is transmitted in a non-persistent manner by several aphid species such as *Acyrthosiphon pisum, Aphis craccivora, Aphis gossypii* and *Myzus persicae.*

Management

1. The virus can also be transmitted by the use of infected cutting tools, so appropriate care should be taken while harvesting.
2. To reduce the sources of initial inoculum, avoid overlapping crops of cucurbits.
3. Plant barrier crops, use yellow sticky traps and spray mineral oils and systemic insecticides.

(B) INSECT-PESTS

1. Red pumpkin beetle (*Aulacophora foveicollis*) (Coleoptera: Chrysomelidae)

Identification

Adults are light orange to dark orange in colour measuring about 0.5-1 cm long. Adult female has pointed tip of abdomen.

Damage

Damage is done by adults, which feed on leaves by biting holes into them. The attack is prominent during May-June. In severe infestation flowers are also damaged and the crops have to be sown again. The larvae (grubs) feed underground plant part. These grubs bore into the roots of the plants.

Management

1. After the crop is over, plough the field deep to kill the grubs present in the soil.
2. Collection and destruction of adult beetles reduce the population.
3. Spray application of *dichlorvos 76%EC (0.05%) or *trichlorfon 5%GR @ 500 grams/ha or *trichlorfon 5% dust @ 500 grams/ha or *trichlorforn 50% EC (0.05%).

   * The use of these insecticides shall be completely banned w.e.f. 31st December, 2020 since these are very toxic to aquatic organism, hence a cautionary statement should be incorporated on label and leaflets that these products are toxic to aquatic organism, hence should not be used near water bodies, aquaculture or pisciculture area. Besides, a cautionary statement should incorporate in label/ leaflet that these products are also toxic to birds.

2. Fruit fly (*Bactrocera tau*) (Diptera: Tephritidae)

Identification

The adults are yellowish brown in colour with lemon yellow curved vertical markings across the thorax. On the apical margin of the wing greyish brown patches are present. Larvae are creamish white and taper anteriorly while the pupa is barrel shape with dull to brown yellow in colour.

Damage and Management

The descriptions about damage and management of fruit flies have been discussed under the insect-pests of tomato.

Beside these, the crop is also attacked by whitefly, aphids, mite, hadda beetle, serpentine leaf miner etc.

Disclaimer

Though utmost care was taken in recommending only approved pesticides for management of different diseases and insect-pests, pesticide recommended wrongly if any, may be purely unintended, inadvertent and by oversight. Further no pesticide can be used in commercial scale without CIBRC approved label claim on the target crop.

7

DISEASES AND INSECT-PESTS OF FRENCH BEAN

Various types of beans are grown throughout the world belonging to different genera of the family Leguminosae. The important members of broad group termed as beans, are common bean or French bean (*Phaseolus vulgaris* L.). French bean is cultivated throughout the world both for green pods and dry pulse. It has gained increasing popularity due to its quality proteins and nutritional balance besides certain medicinal properties. Successful cultivation is hindered due to the attack of various diseases and insect-pests, which reduce the yields drastically. The information on various diseases and insect-pests are described in this chapter.

(A) DISEASES

a) Fungal Diseases

1. Root rot and web blight (*Rhizoctonia solani*)

Symptoms

Appear in following two phaeses.

Root rot

The symptoms of the disease manifest as reddish brown sunken cankers on the root and stem near the soil line which ultimately cause rotting of the root system either partial or complete (Plate 1).

Web blight

Small, circular, water soaked, tan coloured spots appear on leaves (Plate 2a). On pods, dark brown spots appear which are more or less circular, slightly zonated and definitely sunken in shape (Plate 2b). In moist weather, these spots are covered with hyphae and sclerotia of the fungus. Infected seed also show tan brown to reddish brown discolouration.

Pathogen

The disease is caused by *Rhizoctonia solani* (Tel. *Thanatephorus cucumeris*). The pathogen belongs to kingdom Fungi, phylum Basidiomycota, class Basidiomycetes, order Tulasnellales and family Ceratobasidiaceae. The mycelium is septate and branched near the distal septum of the mother hyphal cells at right or acute angles and the branches are constricted at or near the point of origin of the septum. In culture, Sclerotia develop within 5-6 days. The sclerotia consists of brown barrel shaped monilioid cells. The basidial stage of the fungus also occurs but very rarely.

Predisposing factors

The pathogen overwinters either in seed as mycelium or as sclerotia in soil. Root rot development is favoured at high (80%) soil moisture and 25° C temperature while 100 per cent relative humidity coupled with 25° C temperature favoured web blight.

Management

1. Follow long crop rotation.
2. Solarization of the field soil for at least 45 days during summer months.
3. Use healthy seed and treat the seed with carbendazim (0.2%) or captan (0.3%) or *Trichoderma harzianum* (4g/kg seed).
4. Before sowing amend soil with neem cake or mustard cake.
5. To avoid splashing of soil borne inoculums with spattering rain, apply either grass or pine needle mulch on the field floor.
6. Spray the crop with carbendazim (0.1%) and repeat at 10-14 days interval.

2. Anthracnose (*Colletotrichum lindemuthianum*)

Symptoms

On the underside of the leaves, dark brown to black lesions appear along the leaf veins. Circular, reddish-brown to near black sunken lesions with light or gray centers appear on the pods (Plate 3). Seed developing in the infected pods show dark brown to black spots.

Pathogen

The disease is caused by *Colletotrichum lindemuthianum* (Tel. *Glomerella cingulata*). The pathogen belongs to kingdom Fungi, phylum Deutromycota, class Coelomycetes, order Melanconiales and family Melanconiaceae. The mycelium is septate and branched. On the host, stroma below the cuticle is formed. Conidia, which are borne acrogenously on short conidiophores, develop successively under favourable conditions until pinkish spore masses appear on the surface. Setae occur sparingly. Conidia are hyaline, cylindrical with rounded ends with a clear vacuole like body often occurs near the centre. Perithecia contain hyaline and filiform periphyses and asci. Each ascus contains eight ascospores, which may be alantoid in shape. Ascospores are ejected form the tip of ascus.

Predisposing factors

The pathogen overwinters in infected seeds as well as in infected plant debris. Heavy and frequent rains with moderate temperatures (17-25° C) along with high relative humidity (> 90%) favour the development and progress of the disease.

Management

1. Follow 2-3 years crop rotation.
2. Use disease free seed and treat it with carbendazim (0.2%).
3. With the initiation of the disease spray the crop with carbendazim (0.1%). Repeat sprays at 10-14 days interval.

3. Angular leaf spot (*Phaeoisariopsis griseola*)

Symptoms

On leaves, reddish brown angled (3-5) spots appear and such leaves turn yellow (Plate 4a), and fall down prematurely. Superficial, smooth, circular spots with reddish brown centre appear on pods (Plate 4b). Infected seed show yellowish discolouration.

Pathogen

The disease is caused by *Phaeoisariopsis griseola.* The pathogen belongs to kingdom Fungi, phylum Ascomycota, class Deutromycetes, order Moniliales and family Moniliaceae. On the host tissue, the fungus produces a stomatic structure in the substomatal cavity. Conidiophores are laterally joined together in the form of synnemata or coremia, which are darker at the base and lighter towards the tips, and either curved flexuous or septate. Conidia are light grey to hyaline, cylindrical to spindle shaped and straight to slightly curved conidia are borne singly on the tips of columnar hyphae, which are mostly 1-3 (rarely 4-6) septate. No sexual stage of the fungus has yet been observed.

Predisposing factors

The pathogen is both seed and soil borne. Moderate temperatures (24 ± 1° C) coupled with high relative humidity (80%) favour disease development and its progress.

Management

1. Collect and burn the infected plant debris.
2. Follow 2-3 years crop rotation.
3. Use disease free seed and treat the seed with carbendazim (0.2%).
4. With the onset of monsoon rains, spray the crop with carbendazim (0.05%) or hexaconazole (0.05%) and repeat at 10-14 days interval.

4. Floury leaf spot (*Mycovellosiella phaseoli*)

Symptoms

Characteristic symptoms of the disease include the appearance of floury patches on the underside of the leaves followed by necrosis (Plate 5). The corresponding upper surface show brown spots with yellow margins.

Pathogen

The fungus responsible for this disease is *Mycovellosiella phaseoli* (Syn.: *Ramularia phaseolina*). The pathogen belongs to kingdom Fungi, phylum Deutromycota, class Hyphomycetes, order Moniliales and family Moniliaceae. It is an obligate parasite. On the under surface of host, colorless conidiophores emerge in tufts from the stomata and intertwine in clusters around the leaf hairs. Erect, more or less divergent, somewhat tapering towards the tip, simple or irregularly dichotomous, a few short, erect, single conidiophores were also found scattered over the leaf surface. Conidia are mostly ellipsoid or ovoid, tips acute or obtuse, base papillate at the point of attachment, straight hyaline, commonly aseptate, rarely one septate.

Predisposing factors

The pathogen perpetuate in infected plant debris in soil. Moderate temperatures (20° C) along with high humidity favour disease development and its progress.

Management

1. Collect and destroy the infected plant debris.
2. Follow 2–3 years crop rotation.
3. Use healthy seed for sowing.
4. With the onset of monsoon rains, spray the crop with carbendazim (0.05%) and repeat at 10–14 days interval.

5. Rust (*Uromyces appendiculatus*)

The disease appears on all types of beans but it is more severe on pencil type cultivars of beans like Falguni and Payal. The pods of these cultivars are meaty type and tender.

Symptoms

Small yellow or white slightly raised spots appear on either surface of the leaves (Plate 6). These spots enlarge and form reddish brown or rust coloured pustules. A yellow border surrounds these pustules. Severely infected leaves curled upwards, dry up, turned brown and drop down prematurely.

Pathogen

The disease is caused by *Uromyces appendiculatus*. The pathogen(s) belongs to kingdom Fungi, phylum Basidiomycota, class Basidiomycetes, order Uredinales and family Pucciniaceae. It is autoecious rust, generally producing urediniospores and teliospores. The latter may act as a perennation stage. Aecia are rarely seen. Urediniospores serve to disseminate the pathogen over wide areas, they are catenulate, light brown, spiny, one celled and thin walled. They are globoid to ellipsoid in shape. Teliospores are amphigenous, globoid to ellipsoid with a thick wall, generally with a hyaline papilla over the pore and smooth coated. Urediniospores and teliospores can be found in the same sorus.

Predisposing factors

Obligate parasitism and autoecious nature of rust pathogen helps it to complete entire life cycle on bean plants. In cooler regions, survival is ensured through plant debris in the form of teliospores while in other areas through continuous growing of crop including other hosts. Prevalence of cool and warm weather favours its infection.

Management

1. Collect and destroy the infected plant debris.
2. In endemic areas avoid cultivation of pencil beans in monsoon season.
3. Spray the crop with fungicides like hexaconazole (0.1%) or lime sulphur (0.2%) and repeat at 10 days interval.

6. Crown rot (*Sclerotium rolfsii*)

Symptoms

Dark brown, water-soaked lesions appear on the collar region just below the soil line followed by yellowing of leaves and production of a white mouldy growth interspersed with sclerotia at the base of the stem, under surface of the leaves as well as on the pods (Plate 7).

Pathogen

The disease is caused by *Sclerotium rolfsii.* The pathogen belongs to kingdom Fungi, phylum Deutromycota, class Agonomycetes, order Agonomycetales and family unassigned. The mycelium of the fungus is white, coarse and fan shaped. Hyphae are hyaline, branched and septate and characterized by clamp connections. Sclerotia are brown, smooth and round or irregular.

Predisposing factors

The pathogen is soil borne in the form of sclerotia. Disease development and spread is favoured by high soil moisture (30° C) and frequent rains or high soil moisture.

Management

1. Follow long crop rotation.
2. Collect and destroy the infected plant material.
3. With the onset of monsoon rains, spray the crop with captan (0.3%) or hexaconazole (0.05%) and repeat at 10 days interval.

b) Bacterial diseases

7. Common blight (*Xanthomonas axanopodis* pv. *phaseoli*)

Symptoms

Characteristic symptoms include small water soaked spots and yellowing of interveinal areas of leaves (Plate 8). Reddish streaks and cankers appear on stems. On pods, small water soaked spots appear, which are surrounded by distinct zone of narrow reddish brown or brick red band of tissue. On seeds, discoloration appears especially near hilum region.

Pathogen

The bacterium associated with this disease is *Xanthomonas axanopodis* pv. *phaseoli.* The bacterium belongs to kingdom Prokaryotae, division Gracilicutes, class Proteobacteria and family

Pseudomonadaceae. The bacterium is motile, non-capsulated, non-spore forming and G negative with a single polar flagellum. The pathogen forms characteristic yellow coloured colonies due to a water insoluble eacotenoid and a mucoid growth on nutrient glucose agar medium and this yellow colour is due to the production of brominated aryl polyenes, more frequently referred to as xanthomonadins.

Predisposing factors

The bacterium overwinters in infected seed, crop debris and weed hosts.The optimum temperature and relative humidity for disease development are 24° C and 50%, respectively.

Management

1. Collect and destruct the infected plant debris/ weeds.
2. Follow crop rotation.
3. Use disease free seed.
4. Treat the seed with a mixture of Streptocycline (0.01%) + captan (0.25%) for 4 h before sowing.
5. Spray the crop with Streptocycline (0.01-0.015%) and repeat at 7 days interval.

c) Viral diseases

8. Bean common mosaic (Bean common mosaic virus)

Symptoms

The virus causes common mosaic and necrosis (black root) in beans. Common mosaic symptoms appear as leaf rolling or blistering, light and dark-green patches (Plate 9) on the leaf (green mosaic), chlorotic vein banding, yellow mosaic and growth reduction. Systemically infected plants produce few pods which are smaller in size. Infected pods mature late than healthy and may have small, dark-green spots.

Virus

The disease is caused by bean common mosaic virus. The virus belongs to genus *Potyvirus* and family *Potyviridae.* The virus consists of filamentous flexuous particles measuring 730-750 × 12-15 nm in size.

Host range

Besides, French beans, the virus has also been isolated from other leguminous pulse crops like mung bean, braod bean and cowpeas.

Transmission

The virus is seed borne in many legumes. The virus is also transmitted by Aphid vectors like *Acyrthosiphon pisum, Aphis fabae* and *Myzus persicae* in non- persistent manner.

Management

1. Modification of the planting date can reduce the virus incidence in susceptible cultivars.
2. Early plantings of susceptible cultivars at times when the incidence of the aphid vectors is low generally escape the disease.
3. Intercrop bean with maize also help in reducing the incidence of the disease.
4. Always use certified seed to check the disease.
5. Spray application of neem formulation (2-3 ml/lt) also help in reducing the population of aphids.

9. Bean golden mosaic (bean golden mosaic virus)

Symptoms

Symptoms in bean seedlings begin with fine, vein-limited lines of bright yellow chlorosis, usually on the first trifoliate leaves emerging after inoculation. These fine chlorotic lines typically appear first on only half of the leaf near the leaf tips. Veinal chlorosis later expands into a bright golden mosaic that has a striking appearance in the field. Leaves with chlorosis and distortion may become necrotic. Seed setting is very poor if plants are infected at a very early stage.

Virus

The disease is caused by virus bean golden mosaic virus. The virus belongs to the genus *Begomovirus* and family *Geminiviridae*. The virus is characterized by isomeric particles in pairs (geminate),

18-20 nm in dia. Dimmers, 30 nm in length, angular in profile and contains covalently closed circular single stranded DNA- genome.

Host range

The virus is transmitted by whiteflies from infected bean plants to the healthy plants. The alternate host of this virus is *Euphorbia* sp.

Transmission

The virus occurs mostly in tropical and subtropical areas where beans are grown. Bean golden mosaic is transmitted by the whitefly *Bemisia tabaci.* When bean crops are grown in close proximity of potato, cucurbits and tomato having whitefly infection than the incidence of this disease is generally very high in bean crop.

Management

1. Spray the crop with leaf extracts of *Bougainvillea spectabilis, Chenopodium amaranticolor* and *Mirabilis jalapa* on bean plants prior to whitefly appearance help in the reduction of disease incidence.
2. Avoid planting of bean and soybean in the same region is the most practical and effective control measure against the disease.

10. Bean yellow mosaic (bean yellow mosaic virus)

Symptoms

In French bean, symptoms usually appear as mosaic type of leaf variegation with well-defined dark and light-green or chlorotic areas. Prominent islands of green tissue may be more or less elevated/ raised over surrounding chlorotic tissue in which growth and development is reduced so that leaves often become irregularly deformed. Such dark islands occur all over the lamina.

Virus

The disease is caused by bean yellow mosaic virus. The virus belongs to genus *Potyvirus* and family *Potyviridae.* The virus is characterized by elongate, flexuous, filamentous particles about 750-810 × 11-15 nm.

Host range

The virus has wide host range. The major hosts are beans, chickpea, clovers, lentil, peas, soybean and broad bean. The virus also infects crops including groundnuts, cowpea and winged bean.

Transmission

The virus is transmissible in expressed sap and is also seed borne in broad bean. Among the aphid vectors *Aphis craccivora, A. fabae, Acyrthosiphon pisum* and *Myzus persicae* are the most important vectors of the virus.

Management

1. Use disease free seed.
2. Sources of infection may be avoided by spatial isolation of some 100 m, depending on wind direction and vector pressure.
3. Plant perennial crops (such as clovers and gladiolus) near beans.
4. Considerable benefit can be obtained by increasing plant density to achieve a closed crop canopy.
5. Spray application of neem formulation (2–3 ml/lt) also help in reducing the population of aphids.

(B) INSECT– PESTS

1. Black bean bug (*Chauliops nigrescens*) (Hemiptera: Lygaeidae)

Identification

The adults are black in colour about ½ cm long. Hindwings are always membranous. The hardened basal portion is composed of two regions separated by a concave vein Cu^2 -the clavus or narrower area next to the scutellum (when the wings are closed) and the corium. Females are larger than males.

Damage

Damage is caused by both nymphs and adults which insert their mouth parts in the leaves and suck cell sap (Plate 1). They mostly prefer the lower surface of leaves. As a result the vitality of plant is reduced which ultimately affects the quality and the yield is also reduced. Severely attacked leaves show several minute yellowish specks and small, black pustules of excreta. The damaged leaves gradually wither and fall of. The adults appear with the start of rains during the last week of June and remain on the crop until final harvest.

Management

Before flower initiation and pod formation stage, spray the crop with deltamethrin 2.8%EC (0.0025%) or dimethoate 30EC (0.03%).

2. Blister beetles (*Mylabris* spp.) (Coleptera: Meloidae)

Identification

The blister beetles are medium to large size insects. The body is black with three transverse dull orange bands on abdomen. They are soft bodied insects with broad head, which narrows into neck. They are slow flier.

Damage

Damage is caused by adults, which feed on foliage, flowers and developing pods. The larval stages are predatory on eggs of grasshoppers while some eat eggs of solitary bees and their food stores.

Management

Spray application of any one of the following insecticides can be given: cypermethrin 25% EC (0.0075%) or deltamethrin 2.8% EC (0.0028%) at 15 day interval. Besides, neem based formulations can be applied @ 2-3 ml/litre of water.

3. Whitefly (*Trialeurodes vaporariorum*) (Hemiptera: Aleyrodidae)

Description has been given under the insect pests of tomato.

4. Red spider mite (*Tetranychus urticae*) (Acari: Tetranychidae)

Identification

It is generally called as the two spotted red spider mite due to presence of two pigmented spots on the body. The adult mite measures about 0.52 mm in length and 0.30 mm in breadth. The female is oval, pyriform and is variable in colour, sometimes it is ferruginous red, greenish amber or rusty green.

Damage

The damage starts during May–June when the weather is dry. All the active stages of this mite cause the damage by feeding on lower side of leaves. These breed near the petiole and in between the main veins. Under heavy infestation, mites are found all over the leaves. The mite lives gregariously and is protected by silken web. Feeding is accomplished by the chenceral stylets, which penetrate the leaf tissues and the liberated fluid is sucked near the tip of rostrum. Due to extraction of chlorophyll, leaves show characteristic blotching. Due to sucking on lower side of the leaves, uneven growth of two surface occurs which is exhibited by crinkling of leaves from below. The infested leaves show yellowish speckled appearance, which later turn brown. Gradually the infested leaves dry up and a large number of webs are formed on the leaves. The webbing interferes with plant growth. There is a poor setting of pods and yields are considerably reduced.

Management

Biocontrol: The mite is attacked by several natural enemies like coccinellids, thrips, chrysopids, phytoseiid mite etc. Of these bio-agents, Phytoseiid mite, *Amblyseius tetranychivorus* can be mass reared on castor pollen grains and released when the spider mite population appears on the crop.

Chemical control: Same as given under the insect pests of tomato.

Disclaimer

Though utmost care was taken in recommending only approved pesticides for management of different diseases and insect-pests, pesticide recommended wrongly if any, may be purely unintended, inadvertent and by oversight. Further no pesticide can be used in commercial scale without CIBRC approved label claim on the target crop.

◆

8

DISEASES AND INSECT- PESTS OF OKRA

Okra (*Abelmoschus esculentus* (L.) Moench), popularly known as "bhindi" is grown both under open and protected structures. The crop is affected by several diseases and insect-pests which not only reduce the quantity but also quality of the fruits. Diseases and insect-pests of this crop are discussed in this chapter.

(A) DISEASES

a) Fungal Diseases

1. Fusarium wilt (*Fusarium oxysporum* f.sp. *vasinfectum*)

Symptoms

Characteristic symptoms include yellowing and stunting of the plants followed by wilting and rolling of the leaves resulting in the death of the plant (Plate 1). Vascular bundles of the affected plants appear as dark streaks.

Pathogen

The disease is caused by *Fusarium oxysporum* f.sp. *vasinfectum*. The pathogen belongs to kingdom Fungi, phylum Ascomycota, class Deutromycetes, order Moniliales and family Tuberculariaceae. The mycelium is hyaline septate and intercellular in the host. Microconidia (0-1septate) and macroconidia (3-5 septate) are produced on sporodochia and pinnotes. In mass these conidia appear buff to salmon orange in colour. Both intercalary and terminal chlamydospores are formed which are broadly ovate.

Predisposing factors

The pathogen perpetuate in the infected seed and soil. Moisture is not important for infection but high temperature is essential. Optimum temperature for disease development is 22–28° C.

Management

1. Follow long crop rotation.
2. Soil solarization during summer months for at least 40 days.
3. Collect and destroy roots of the infected plants.
4. Use resistant cvs. like Pusa Sawani and Pusa Makhmali.
5. Use healthy seed.

2. Powdery mildew (*Erysiphe cichoracearum*)

Symptoms

The symptoms of the disease appear as minute discoloured patches with thin fine meshwork of white mycelium arising at many places on the upper surface of the lower most leaves (Plate 2). Soon these white patches join together to form larger white coating on both surfaces of the leaves. Affected leaves later turn yellow and drop down prematurely.

Pathogen

The fungus responsible for this disease is *Erysiphe cichoracearum*. The pathogen belongs to kingdom Fungi, phylum Ascomycota, class Ascomycetes, order Erysiphales and family Erysiphaceae. The conidia are single celled, hyaline, barrel-shaped and in long chains. Cleistothecia are globose, dark, with hyaline to dark brown myceloid appendages. They contain 18–80 asci. Each ascus usually contains two asocospores but very rarely three. Ascospores are one celled, hyaline and oval to subcylindrical in shape.

Predisposing factors

Wherever clesitothecia are formed they explain the disease cycle. Otherwise the fungus overwinters in conidial stage either in protected structures or on wild hosts. Moderate temperatures, reduced light, fertile dry soil and succulent plant growth favour the disease development.

Management

1. Collect and destruct the infected plant parts regularly.
2. Spray the crop with wettable sulphur (0.25%) and repeat at 10–14 days interval.

3. Cercospora leaf spot (*Cercospora* spp.)

Symptoms

Three different species of *Cercospora* have been found associated with this disease.

1. ***Cercospora abelmoschi:*** The fungus causes no definite leaf spot but on lower surface of the leaves grows as a sooty mould (Plate 3). Severely affected leaves roll, wilt and fall down.
2. ***Cercospora hibiscina:*** The fungus produces dark olivaceous patches of mould on the lower surface of the leaves.
3. ***Cercospora malayensis:*** Definite lesions appear on the leaves with grey centers and red to purple borders.

Pathogen

The following three species of *Cercospora* are associated with this disease: The pathogen belongs to kingdom Fungi, phylum Ascomycota, class Deutromycetes, order Moniliales and family Dematiaceae.

Cercospora abelmoschi: Conidiophores are long, brown and bear pale olivaceous, slightly tapered conidia.

Cercospora hibiscina: Conidiophores are extremely long sometimes up to 1000 µm in size, narrow and bear conidia that are sometimes hyaline and appreciably narrower than *C. abelmoschi.*

Cercospora malayensis: Conidiophores are borne in clusters of 5–20 and bear conidia, which are colourless, narrow, long and tapering from the blunt base to the sharp tip.

Predisposing factors

These fungi overwinter in the infected plant debris and also on wild hosts. Moderate temperatures and high humidity favour disease development.

Management

1. Collect and destruct the infected plant debris.
2. Follow crop rotation.
3. Spray the crop with Bordeaux mixture (4:4:50) and repeat at 10-14 days interval.

b) Viral diseases

4. Yellow vein mosaic (yellow vein mosaic virus)

Symptoms

The characteristic symptoms are clearing of small veins and veinal chlorosis of leaves. On leaves, yellow network of the veins is very conspicuous and veins and vein lets are thickened (Plate 4). The chlorosis extends to the interveinal areas and whole leaf turns yellow in case of severe infection. The fruits are dwarfed, malformed and yellowish green in colour. The infected fruits fetch poor prices in the market.

Virus

The disease is caused by yellow vein mosaic virus. Virus particles are spherical, isometric, measuring 28-30 nm in diameter.

Transmission

The virus is not sap transmissible but under artificial conditions it can be transmitted by grafting. In nature the virus is transmitted by insect vector, white fly (*Bemisia tabaci* Genn.).

Host range

Host range of this virus includes *Croton sparsiflora, Malvastrum tricuspidatum, Abelmoschus manihot, Althaea rosea, Hibiscus tetraphyllus* and *Ageratum* sp.

Management

1. Destruct wild hosts from in and around the field.

2. Avoid mixed cropping of pumpkin with okra.
3. Avoid period of maximum population of the insect vector whitefly by adjusting date of sowing and remove affected plants regularly up to 55 days.
4. Application of yellow coloured polythene mulch on field floor significantly delayed symptom appearance.
5. Use resistant varieties like Punjab, Punjab Padmini, Punjab-8, Hisar Unnat and Azad Bhindi-1 for cultivation.
6. Sprays of systemic insecticide(s) or neem formulations (2-3ml/lt) can keep the population of vector in check.

(B) INSECT- PESTS

1. Blister beetle (*Mylabris* spp.) (Coleoptera: Meloidae)

Identification

Same as described under the insect-pests of French bean.

Damage

The damage is done by adult beetles, which feed on flowers due to which fruit is not formed.

Management

Same as described under the insect-pests of French bean.

2. Cotton aphid (*Aphis gossypii*) (Hemiptera: Aphididae)

Identification

The general body colour is greenish brown.

Damage

Both adults and nymphs are the damaging stages which insert their mouth parts inside leaves and suck sap. More population of this pest results in curling of leaves which further start drying. The plant remains stunted. Besides causing direct loss, it is also capable of transmitting viral diseases. Heavy population of this aphid leaves honeydew on leaves on which sooty mould develops which interferes with photosynthetic activity of the plant and thus yield is reduced.

Management

1. Spray application of any one of the following insecticides:

   Acetamiprid 20%SP (0.01%) or dimethoate 30%EC (0.03%) or imidacloprid 17.8%SL (0.15%).

3. Red cotton bug (*Dysdercus koengii*) (Hemiptera: Pyrrhocoridae)

Identification

The adult is mediumsized reddish bug with white ba nds on the abdomen and black markings on wings. The antennae, scutellum and the membranous portion of the forewings are black.

Damage

Both adults and nymphs cause damage by sucking sap from leaves and fruits. Due to continuous sucking of sap, the plant weakens which ultimately affects the yield. In severe infestation, the curling of fruits takes place.

Management

1. Hand picking and killing of insects in kerosinized water is effective in checking the population of this pest.
2. Deep ploughing during summer months exposes the eggs thus causing their desiccation.
3. The following insecticides can be used against this pest:

   Fluvalinate 25% EC (0.01%) or neem based formulation @ 2-3 ml/litre of water.

4. Fruit borer (*Helicoverpa armigera*) (Lepidoptera: Noctuidae)

Identification

Same as described under tomato.

Damage

Damage is caused by larva which borers into tender shoots, moves downwards, thus block the passage for movement of nutrient. The infested shoots ultimately wither, thus growing point is killed. The larvae then start entering bud and fruits of okra causing heavy shedding of fruiting bodies. Fruits become distorted. The infested fruits are contaminated with insect excreta, which is not suitable for human consumption and marketing.

Management

Same as described under the insect-pests of tomato.

5. Chrosymelid beetle (*Nisotra chrysomeloides*) (Coleoptera: Chrysomelidae)

Identification

The female adults are brownish black while males are somewhat lighter in colour. Females are bigger than males and are creamish and spindle shaped.

Damage

Adult beetle does the damage by making round holes on the leaves, which get skeletonized under heavy infestation. On leaves, mostly the beetles feed on upper surface. Sometimes flowers and small fruits are attacked. Younger leaves are preferred over the older ones.

Management

1. Neem based formulations @ 2–3 ml/litre of water.
2. Neem seed kernel extract (4%).

6. Jassid (*Amrasca biguttula biguttula*) (Hemiptera: Cicadellidae)

Identification

The adults are about 4 mm in length with light yellow colour body having greenish tinge. On slight disturbance the nymphs move briskly whereas the adults fly away in all directions.

Damage

Damage is caused by adults and nymphs by sucking cell sap from lower parts of leaves. Due to sap sucking, leaves turn yellow near the margin and show upward cupping. In severe infestation, the plant shows "hopper burn". The population is maximum during summers, however, during monsoon the population declines.

Management

1. Application of following insecticides may be used to manage the population of this pest:

   Fenvalerate 20%EC (0.01%) or imidacloprid 17.8%SL (0.01%) or lambda cyhalothrin 5%EC (0.005%) or thiomethoxam 25%WG (0.025%).

Disclaimer

Though utmost care was taken in recommending only approved pesticides for management of different diseases and insect-pests, pesticide recommended wrongly if any, may be purely unintended, inadvertent and by oversight. Further no pesticide can be used in commercial scale without CIBRC approved label claim on the target crop.

✦

9

DISEASES AND INSECT- PESTS OF PEA

Among leguminosae family, Pea (*Pisum sativum* L.) is an important crop, which includes both ornamental and edible forms, the garden pea and field pea and grown over a greater part of the world. During cultivation, the crop is attacked by several diseases and insect-pests, which not only reduce the yields but also impair the quality of the produce. A comprehensive account of various diseases and insect-pests affecting this crop are given herein along with their management practices.

(A) DISEASES

a) Fungal diseases

1. Powdery mildew (*Erysiphe pisi*)

Symptoms

White floury patches covering large areas appear on all above green parts of the plants like leaves, tendrils, stems and pods (Plate 1). On pods, these patches first turn light brown and finally dark brown in colour and become necrotic.

Pathogen

The disease is caused by *Erysiphe pisi*. The fungus is obligate ectoparasite, spreading on the surface of the host and sending haustoria into the epidermal cells to draw nourishment. The conidiophores are spetate. The conidia are ovate, barrel shaped, hyaline and unicellular. Sexual fruiting bodies, cleistothecia appear as dark coloured, round structures, sharp, minute bodies scattered in the mycelial web. Each cleistothecium contains usually 2-8 asci. Each ascus contains 3-8 ascospores.

Predisposing factors

The pathogen is an obligate parasite and perpetuate either through cleistothecia (in dry temperate regions) or in the form of mycelium and conidia on the host and on annual and perennial both cultivated and wild hosts. The disease is more prevalent in dry weather and moderate temperatures. A temperature range of 10-30° C is favourable for conidia germination with optimum being at 20-24° C.

Management

1. Early sown (late October) escapes the disease.
2. Avoid close planting.
3. Use resistant varieties.
4. With the initiation of the disease, spray the crop either with wettable sulphur (0.2%) or carbendazim (0.05%) or triadimefon (0.1%) and repeat at 10-14 days interval.

2. Fusarium Wilt (*Fusarium oxysporum* f.sp. *pisi*)

Symptoms

Symptoms appear as yellowing of the lower leaves and a stunting or dwarfing of plant growth (Plate 2). The stipules and margins of the leaflets curl downward and inward. Xylem tissues of infected plants show black discolouration. Severely infected plants may wilt and die.

Pathogen

The fungus responsible for this disease is *Fusarium oxysporum* f.sp. *pisi.* The mycelium is septate, microconidia are generally one celled and macroconidia are fusiform, hyaline and multiseptate. Infavourable weather conditions, chlamydospores are also produced.

Predisposing factors

The pathogen is both seed and soil borne. Higher soil temperature ranging between 21.8 to 22.5° C and soil moisture in between 13.0 to 14.4% favour disease development.

Management

1. Follow at least 3 years crop rotation and solarize field soil for 45 days before sowing.
2. Avoid early sowing of crop in infested areas.
3. Use balanced doses of fertilizers.
4. Use resistant cultivars, if available.
5. Use healthy seed and dip the seed in carbendazim (0.1%) solution for 10 minutes or treat the seed with *Trichoderma harzianum* (4g/kg seed) formulation.

3. White Rot (*Sclerotinia sclerotiorum*)

Symptoms

The infection may occur at any part of the foliage, but occurs mainly on the stems or branches as dry, discoloured spot (Plate 3a). Pods are also infected and flesh start rotting (Plate 3b). In moist weather affected tissue is covered with fluffy, white mycelium. Prominent dark sclerotia, irregular in size and shape are found on the infected plant surface as well as on pods (Plate 3b).

Pathogen

The disease is caused by *Sclerotinia sclerotiorum*. The mycelium is hyaline, septate and branched. The mycelium forms sclerotia. Apothecia emerge through pores in rind of the sclerotia when moisture is abundant. Apothecia are tan, fleshy, funnel shaped cups on thin stalks. Ascospores are discharged from the apothecia in abundance.

Predisposing factors

The pathogen perpetuates in the form of sclerotia left in soil or plant debris. Low temperature and high moisture on the plant surface favour the initiation of disease in the post flowering stages of the crop.

Management

1. Collect and destroy all infected plant debris.
2. Follow crop rotation with paddy.
3. Use resistant sources if available.

4. Maintain wider row spacing (50-60 cm).
5. Apply *Trichoderma harzianum* formulation at the time of sowing.
6. Apply foliar sprays of carbendazim (0.1%) at flowering stage and repeat at 10-14 days interval.

4. Downy Mildew (*Peronospora pisi*)

Symptoms

Yellow to brown spots appear on the upper surface of leaves, while the corresponding under surface is covered with a grayish violet downy growth of the fungus. On pods, the patches are pale green, more or less elliptical which later turns brown in colour.

Pathogen

The fungus responsible for this disease is *Peronospora pisi.* The pathogen belongs to kingdom Chromista, phylum Oomycota, class Oomycetes, order Peronosporales and family Peronosporaceae. The fungus is an obligate parasite. Mycelium is hyaline, intercellular and branched with finger shaped haustoria. In favourable weather conditions sporangiophores and sporangia are produced on the under surface of the leaves. Oospores are also produced in unfavourable weather conditions.

Predisposing factors

The pathogen perpetuate in soil through oospores. Sometimes, it is seed borne also either under the seed coat or as a contaminant with seed lots. Disease occurrence is usually associated with unusually wet weather and long periods of cool overcast weather.

Management

1. Collect and destroy the infected plant debris.
2. Follow long crop rotation.
3. Use healthy seed.
4. Maintain proper drainage of the field.
5. With the initiation of the disease, spray the crop with Bordeaux Mixture 4:4:50 and repeat at 10-14 days interval.

5. Rust (*Uromyces pisi* and *U. viciae fabae*)

Symptoms

On leaves symptoms of the disease appear as minute, raised pustules, which later become distinct, yellowish, circular sori (Plate 4a). The yellow spots bear aecia and in same sorus uredo pustules develop on leaves as well as on other parts of the plants. Later in the season, teleutopustules develop on stems (Plate 4b) and rarely on pods. Later on these pustules turn dark brown to black in colour.

Pathogen

The disease is caused by two species of *Uromyces* viz., *U. pisi* and *U. viciae fabae*. The pathogen belongs to kingdom Fungi, phylum Basidiomycota, class Basidiomycetes, order Uredinales and family Pucciniaceae. It is autoecious rust. The fungus is heterothallic. Pycnia occur in small groups. Aecia are cupulate and aeciospores are round to angular or elliptical with hyaline wall. The uredia present a powdery appearance. Urediniospores are round to ovate, light brown and echinulate. The telia are dark brown to black. Teliospores are subglobose, ovate to elliptic with round or flattened apex which is considerably thickened and appears papillate.

Predisposing factors

Pea rust is mainly soil borne in the form of teliospores in the infected plant debris. The rust also survives on weed hosts. The prevalence of low temperature (17-22° C) coupled with high humidity, dew or frequent light rain showers are favourable for initiation of the disease by aeciospores. As the temperature increases i.e. 18-25° C coupled with rain or dew, the disease attains the epiphytotic proportions.

Management

1. Destruction of diseased crop debris.
2. Follow long crop rotation. Avoid broad beans, *Vicia* and *Lathyrus* in rotation.
3. Sowing of the seed should be done early in disease prone areas.

4. Use resistant varieties.
5. With initiation of the disease, spray the crop with wettable sulphur (0.2%) or triadimefon (0.1%) and repeat at 10-14 days interval.

6. Ascochyta Foot Rot and Blight (*Ascochyta pinodes, Phoma medicaginis* var. *pinodella* syn. *Ascochyta pinodella, A. pisi*)

Symptoms

Ascochyta or Mycosphaerella blight: Small purple spots appear on leaves, which later turn brown to black in colour and become zonated (Plate 5). Lesions on stems are elongated and brown or black in colour.

Ascochyta foot rot: Dark brown lesions appear on stem soon after seed germination, extending upward on the stem and downward towards roots (Plate 6).

Ascochyta leaf and pod spot: The leaf spots are large, pale brown in colour and inclined to be papery having tan centre (Plate 7).

Pathogen(s)

Different species of *Ascochyta* are known to cause this disease. The pathogen (s) belong to kingdom *Fungi*, phylum *Ascomycota*, class *Deutromycetes*, order *Sphaeropsidales* and family *Sphaeropsidaceae*.

Ascochyta pinnodes: produces perithecial stage *Mycosphaerella pinodes*. Perithecia are globose having beaked ostioles and contain bitunicate, cylindrical, clavate asci.Each ascus contains eight ascospores which are bicelled and hyline. The conidia have large number of globules. The spore ooze collectively appears buff coloured.

Ascochyta pinodella (Phoma medicaginis* var. *pinodella): produces single sometimes bicelled hyaline spores in pycnidia. No perfect stage known.

Ascochyta pisi: Conidia are longer and thin with slight constriction at septation.

Predisposing factors

The pathogen(s) perpetuate both in the soil as well as in infected seed. Frequent rains along with moderate temperatures (20° C) help in buildup of epidemic. High humidity i.e. 80% or more is essential for infection.

Management

1. Collect and destruct diseased plant debris.
2. Follow 3-4 years crop rotation.
3. Use healthy seed produced in low rainfall areas.
4. Treat the seed with captan (0.3%) or carbendazim (0.2%).
5. Though the sprayings are not very economical but carbendazim (0.1%) can check the disease to some extent.

b) Bacterial diseases

7. Bacterial Blight (*Pseudomonas syringae* pv. *syringae*, *P. syringae* pv.*pisi*)

Symptoms

All aerial parts of the plants are infected by the disease. Brown water soaked spots appear on the leaf margins, nodal points and pods (Plate 8). If infection takes place at early stages of plant growth, results in complete wilting of plants.

Pathogen

Two species of *Pseudomonas* like *P. syringae* pv. *syringae* and *P. syringae* pv. *pisi* are known to cause this disease. The bacteria belong to kingdom *Prokaryotae,* division *Gracilicutes,* class *Proteobacteria* and family *Pseudomonadaceae.* The bacterium is Gram negative, non-spore forming and motile rods with one or more polar flagella.

Predisposing factors

The bacterium perpetuate in the infected seed. Cool and moist weather favour disease development. Frost is an important factor affecting disease development. Wet soil favours bacterium transmission from the infected seed to resulting plant.

Management

1. Collect and burn the infected crop debris.
2. Follow 3 years crop rotation.
3. Use disease free seed.
4. With the initiation of the disease spray the crop with Bordeaux mixture (4:4:50) and repeat at 10 days interval.

c) Viral diseases

8. Pea seed-borne mosaic (pea seed-borne mosaic virus)

Symptoms

Symptoms on peas can be influenced by plant age at the time of infection, temperature and pea genotype. Symptoms of the disease include leaf chlorosis, whole plant stunting, shortening and downward rolling of the leaflets, vein clearing and swelling, apical malformation (rosetting), mosaic, distorted flowers or seed pods and failure to set pod.

Virus

The disease is caused by pea seed borne virus. The virus responsible for this disease belongs to genus *Potyvirus* and family *Potyviridae*. The virus is a 780 nm flexuous particle with ss-RNA genome.

Host range

The virus has a wide host range but there are only three significant hosts like peas, lentils and broad bean which are economically important as far as seed borne dissemination and crop loss is concerned.

Transmission

The virus is transmitted mechanically, by seed transmission and by aphid vectors, either semi-persistently or non-persistently. The natural vectors are *Acyrthosiphon pisum, Aphis craccivora* and *A. fabae*.

Management

1. Use cultivars that are resistant to the virus.
2. Since seed borne viruses create a within-field inoculum

source, sprays of any systemic insecticide or neem formulation (2-3 ml/lt) can reduce vector population which will ultimately check virus spread.

3. Reflective mulches/ yellow sticky traps are also used to keep the incidence of the disease in check.

9. Pea early-browning (pea early browning virus)

Symptoms

Symptoms in the crop appear as brown patches of poor growth, particularly when infection occurs early in the season. Plants within these patches are retarded and may die. Stems may break easily. Pods can show purple discoloration and necrotic flecks, ring patterns or pits, and may be small and distorted following early infection.

Virus

The disease is caused by pea early browning virus. The virus belongs to genus *Tobravirus* and family *Unassigned.*

Host range

Natural infection with this virus occurs mainly in legume species, including pea, broad bean and French bean. Some leguminous and weed species could be naturally infected in the field.

Transmission

The virus is transmitted in soil by species of freeliving plant parasitic nematodes of the family *Trichodoridae*, specifically by *Trichodorus* and *Paratrichodorus* species. These nematodes prefer free-draining sandy or light clay soils. The virus is also seed borne in several hosts.

Management

1. Use disease free seed.
2. Weed control may also help in reducing sources of infection, and persistence of virus in the vector population.
3. Chemical treatments, which kill nematodes, or inhibit their activity, can be applied as soil fumigants, which become systemic in the host plants.

(B) INSECT-PESTS

1. Pea leafminer (*Chromatomyia horticola*) (Diptera: Agromyzidae)

Identification

The adults are small flies with two wings (one pair of wing). The body is grayish black and the head is light brown. The pupa is fusiform with distinctly defined segments.

Damage

The adults with their sucking and sponging types of mouthparts feed on sap of the plants. The female makes punctures on the leaves with the help of ovipositor and then feeds on the emerging sap. The males cannot make such punctures. The pale-yellow specks caused at the site of punctures, may lateron change into prominent swelling due to increased activity of the injured tissues. The larvae cause more serious damage. They make prominent whitish tunnels in the leaves. Excess tunnels interfere with proper photosynthesis. The manufacture of food by the leaves is thus adversely affected. In severe infestation, the attacked leaves wither and die. Consequently the flowering and the fruiting of the infested crop is much reduced.

Management

Biological Control: The population of the pea leaf miner is naturally kept under check by a large number of larval and pupal parasitoids. These include Braconids (*Bacnusa* sp.) and Eulophids (*Chrysonotomyia* sp.; *Diglyphus* sp.; *Pediobius* sp.).

Chemical Control: Application of oxydemeton methyl 25%EC (0.025%) or *dichlorvos (0.04%). When spraying coincides with pod formation, harvest the crop before spraying. If the attack persists, the spray of the above insecticides can be repeated.

* The use of this insecticide shall be completely banned w.e.f. 31st December, 2020 since it is very toxic to aquatic organism, hence a cautionary statement should be incorporated on label and leaflets that these products are toxic to aquatic organism, hence should not be used near water bodies, aquaculture or pisciculture area.

2. Pea pod borer (*Helicoverpa armigera*) (Lepidotera: Noctuidae)

Identification

The moth is stoutly built. On the apical margin of forewing, wavy lines in the form of light black bands are visible. The hindwings are whitish and lighter in colour with a broad blackish band along the outer margin. The pupa is broadly rounded anteriorly but tapers posteriorly.

Damage

The damage is done by larvae, which in the initial stage feed on pea leaves but in the later instars they feed inside pea pod, thus cause reduction in yield.

Management

Same as for the control of tomato fruit borer.

3. Semilooper (*Thysanoplusia orichalcea*) (Lepidoptera: Noctuidae)

Identification

The forewings of adults are light brown from centre towards outer margin of wings on dorsal side and with a larger golden patch. Before population, the larva spins a cocoon, which is broadly rounded anteriorly but tapers posteriorly.

Damage

This is a polyphagus pest of varying severity on a number of crops. Larvae cause the damage which in the earlier instars (I & II instars) scrap the leaf surface while the third, fourth, and fifth instars larvae consume the whole leaf.

Management

1. Same as for the control of cabbage semilooper under Cruciferous crops.

4. Pea blue butterfly (*Lampides boeticus*) (Lepidoptera: Lycaenidae)

Identification

The adult butterfly is violet in colour with bluish tinge. There is a sexual dimorphism. The males have blue violet upper face whereas in female, the blue colour is concentrated in the centre of the wing.

Damage

Larva causes the damage which bores into floral buds, flowers and pods. In the developing pods the grains yield is reduced drastically.

Management

Spray application of deltamethrin 2.8EC (0.0025%) at the time of bud formation stage.

5. Pea Thrips (*Caliothrips indicus*) (Thysanoptera: Thripidae)

Identification

The adults are very small about 0.7 to 0.9 mm in length. The males are wingless where the females have long, narrow wings with long hair along the hind margins. The nymphs are wingless and are smaller than adults but resemble the adults in shape and colour.

Damage

They have asymmetrical piercing mouthparts. Damage is caused by nymphs and adults which feed on sap which is oozed out by scratching leaf surface with their piercing mouthparts due to which white patches on the affected leaves develop. Since the pest is very active during flowering, both yields and viability of the seeds are reduced.

Management

1. Spray application of acetamiprid 20%SP (0.01%) or chlorpyriphos 20%EC (0.02%). Applications of neem based formulations also help in reducing the population of this pest.

Disclaimer

Though utmost care was taken in recommending only approved pesticides for management of different diseases and insect-pests, pesticide recommended wrongly if any, may be purely unintended, inadvertent and by oversight. Further no pesticide can be used in commercial scale without CIBRC approved label claim on the target crop.

10

DISEASES AND INSECT-PESTS OF CRUCIFERS

The cruciferous vegetables belong to family cruciferae and most of which fall in the Genus *Brassica*. Generally the cruciferous vegetables are grown in cool weather and almost all the parts of these plants are eaten in one or other form. Because most of the vegetables belong to the same family, so the diseases and insect pests attacking these vegetables are same and the details of these are given as under.

(A) DISEASES

a) Fungal diseases

1. Stalk Rot (*Sclerotinia sclerotiorum*)

This is an important disease of seed crop particularly of late (Snowball) group of varieties.

Symptoms

Leaves of the infected plants lose their shine and droop down on the ground (Plate 1a). Stalk of the infected plants are girdled at soil line and start rotting from inside. Ultimately such infected stalks become hollow and are filled with black sclerotia of the pathogen (Plate 1b). Curds of the infected plants also become loose and show characteristic white rot symptoms. In cool moist weather, watery soft lesions also develop on bolting shoots which later exhibit white silvery appearance followed by complete wilting. There is no pod formation in such plants.

Pathogen

The disease is caused by *Sclerotinia sclerotiorum*. The pathogen belongs to kingdom Fungi, phylum Ascomycota, class Discomycetes, order Helotiales and family Sclerotiniaceae.

Mycelium is white, cottony, branched, consisting of closely septate hyphae. Micro-conidia are produced on short lateral branches of the vegetative mycelium in chains. When the food supply is exhausted and the vegetative growth ceases, the hyphae collect in small dense masses and form sclerotia. These sclerotia are first white in colour and then turn black. Apothecia are tan, fleshy, funnel-shaped cups on thin stalks. Ascospores are discharged from the apothecia in abundance.

Predisposing factors

The pathogen survives in soil as sclerotia. Cool (15.5-21° C) and moist weather along with high soil moisture favour disease development and its progress.

Management

1. Follow cauliflower-paddy-cauliflower or cauliflower-maize-cauliflower rotation.
2. Remove affected leaves at weekly interval and destroy them.
3. Apply pine needle or sunflower inflorescence residue mulches on the field floor.
4. Application of decomposable organic matters like sunflower or mustard cakes and bean straw also help in reducing the incidence of this disease.
5. Soil application of antagonists like *Trichoderma harzianum* or *T.viride* also proved effective in controlling this disease, if applied at the time of earthing up.
6. Spray the crop with zineb 75 WP (0.25%). Initiate sprays at earthing up stage and continue at 10 to 14 days interval till bolting stage.

2. Curd rot Complex (*Fusarium* spp., *Sclerotinia sclerotiorum*, *Alternaria* spp., *Erwinia caratovora* pv. *caratovora*)

Symptoms

Curds of the infected plants start rotting from any portion (Plate 2). The rotting usually starts after some injury caused by hailing/ frost or due to deficiency of boron.

Pathogen

This is a complex disease and caused by fungi and bacteria. The pathogens responsible for this disease are *Fusarium* spp., *Sclerotinia sclerotiorum, Alternaria* spp., *Erwinia caratovora* pv. caratovora. The bacterium belongs to Kingdom Prokaryotae, division Gracilicutes, class Proteobacteria and family Enterobacteriaceae. Bacterium is Gram negative. Other pathogens are described in this chapter under different diseases.

Predisposing factors

Pathogens responsible for this disease mostly overwinter in infected plant debris/soil under field conditions as mycelium, chlamydospore and sclerotia while bacterium is omnipresent. The pathogens either enter through the natural openings or through the injury caused by extreme environmental conditions like hailing or frost. High humidity and moderate temperatures favour disease development.

Management

1. Protective sprays with a mixture of Zineb 75WP (0.25%) or Bordeaux mixture (4:4:50) may be given to curds particularly before there is likelihood of frost. Repeat twice after 8–10 days interval.
2. Small rotten areas on the curds may be cut with knife and painted with 4:4:50 Bordeaux mixture.

3. Downy Mildew (*Peronospora parasitica*)

Symptoms

The symptoms appear as angular translucid spots, which are purplish brown on the underside while corresponding upper surface is tan to yellow in colour (Plate 3a). It also causes curd rot. Rotten tissues become brown and are bordered with black discolouration at the base of the curd stalk (Plage 3b).

Pathogen

The fungus responsible for this disease is *Peronospora parasitica* which is an obligate parasite. The pathogen belongs to kingdom Chromista, phylum Oomycota, class Oomycetes, order Peronosporales and family Peronosporaceae. The mycelium is intrercellular with large, finger–shaped intracellular haustoria. The dichotomously branched conidiophores are produced during darkness, which ultimately bifurcate to form sterigmata which bear the single conidium at the tip. The conidia are broadly oval, ellipsoidal and hyaline. The oospores are globose and appear pale yellow in colour and germinate by germtube.

Predisposing factors

The pathogen survives as oospores in senescence host tissues and as latent systemic mycelium in seeds or in infected plant debris. The disease is favoured by cool and moist weather conditions.

Management

1. Collect and destroy infected plant debris.
2. Remove cruciferous weeds around the field.
3. Use clean/ disease free seed (without fragments of pods which may contain oospores).
4. Treat the seed with Apron (0.3%).
5. Spray the crop with azoxystrobin +chlorothalonil (0.2%) or zineb 75WP (0.25%) at 10–14 days interval, first spray being just at the appearance of disease.

4. Alternaria Leaf Spots (*Alternaria brassicae, A. brassicicola* and *A. raphani*)

Symptoms

On leaves, symptoms appear as minute dark brown to black spots, which may enlarge by forming concentric rings and are surrounded by a yellow halo (Plate 4). In humid weather fungal spores may appear as a bluish growth in the centre of spots. Browning of cauliflower heads also occur.

Pathogen(s)

Three species of *Alternaria* are responsible for this disease. *A. brassicicola* and *A. brassicae* attack most of the crucifer vegetables while *A. raphani* usually attacks radish. The pathogen(s) belongs to kingdom Fungi, phylum Ascomycota, class Deutromycetes, order Moniliales and family Dematiaceae.

A. brassicae: Mycelium is immersed, hyphae branched, septate, hyaline. Conidiophores are simple, erect, bearing one to several small but distinct conidial scars. Conidia are solitary or occasionally in chains of up to 4, obclavate, rostrate with 16-19 transverse septa and 0-8 longitudinal or oblique septa, pale or greyish olive, the beak about 1/3 to 1/2 the length of the conidium.

A. brassicicola: Mycelium immersed, hyphae branched septate, hyaline at first, later brown or olivaceous brown. Conidiophores arising through stomata either singly or in groups of 2-12, usually simple and erect conidia mostly in chains of up to 20 or more, nearly cylindrical, usually tapering slightly towards the apex or obclavate, the basal cell rounded, the beak usually almost non-existent, mostly less than 6, transverse septa, often constricted at the septa, pale to dark olivaceous brown, smooth or becoming slightly warted with age.

A. raphani: Conidiophores are simple or occasionally branched, septate and olivaceous brown in colour. Conidia are formed commonly in chains of 2-3, straight or slightly curved, obclavate or ellipsoidal, generally with a short beak, mid to dark golden brown or olivaceous brown in colour, smooth or sometimes minutely verruculose, with 3-7 transverse and often a number of longitudinal or oblique septa.

Predisposing factors

The pathogen(s) overwinter in infected plant debris, in or on infected seed and cruciferous weeds. Moderate temperatures (21-27° C), high humidity and splashing rains favour development of these diseases.

Management

1. Collection and destruction of infected plant debris.
2. Use disease free seed.
3. Treat the seed with captan (0.3%).
4. Spray the crop with Zineb 75WP (0.25%) or tebuconazole (0.06%) or azoxystrobin + chlorothalonil (0.2%) andrepeat at 10-14 days interval.

5. White rust (*Albugo candida*)

Symptoms

The symptoms of the disease appear as isolated pustules on the leaves and stems which show characteristic raised glossy blisters of 1-2 mm in diameter (Plate 5). Blisters after rupturing the host epidermis, release white powdery mass comprising of sporangia on the host surface. Systemic infection stimulates hypertrophy and hyperplasia of stem and inflorescence.

Pathogen

The disease is caused by *Albugo candida*. It is an obligate parasite. The pathogen belongs to kingdom Chromista, phylum Oomycota, class Oomycetes, order Peronosporales and family Albuginaceae. Mycelium is intercellular and non-septate. Numerous short conidiophores arise from the mycelium, which are arranged in rows beneath the host epidermal layer press on the epidermis to cause the pustule/ blister. Inside the pustules, sporangia are produced in chains arising from the sporangiophores at the base of the cavity. The sporangia are hyaline, spherical, thick walled and hyaline. They germinate either by germtube or by zoospores. The oospore is spherical with a thick irregular wall, deep yellow to dark brown in colour.

Predisposing factors

The fungus subsists from one season to another in the form of oospores and perennial mycelium in cruciferous weeds. Cool and moist weather favours the disease development. Optimum temperature for germination of sporangia through zoospores is 20° C and moisture is essential for sporangial germination.

Management

1. Collect and burn the infected plant debris.
2. Follow crop rotation.
3. Avoid application of excess nitrogen.
4. Use resistant cv./ hybrid, if available.
5. Treat the seed with captan (0.3%).
6. With the initiation of the disease, spray the crop with azoxystrobin + chlorothalonil (0.2%) or zineb (0.25%)

followed by sprays of Bordeaux mixture (4:4:50) and repeat at 10 days interval.

b) Bacterial diseases

6. Black Rot (*Xanthomonas campestris* pv. *campestris*)

Symptoms

The symptoms of the disease appear as 'V' shaped lesions with its base towards the mid-rib and the infected tissues turn yellow and chlorotic (Plate 6). The veins and veinlets of the chlorotic tissues become dark brown to black in colour. Due to infection, heads become brown to black. Later on the infected tissues start rotting.

Pathogen

The disease is caused by *Xanthomonas campestris* pv. *campestris*. The bacterium belongs to Kingdom Prokaryotae, division Gracilicutes, class Proteobacteria and family Pseudomonadeaceae. It is a small, rod shaped, aerobic, Gram negative, non-spore forming bacterium with single polar flagellum. It is catalase positive, hydrogen sulphide positive, oxidase negative and does not produce nitrate or indole. It produces a yellowish extracellular polysaccharide (EPS) called Xantham on media containing glucose.

Predisposing factors

The bacterium overwinters in infected seed, diseased plant debris and on cruciferous weeds. Optimum temperature for the disease development is about 26.5 to 30° C. Wet warm weather favours disease development. Heavy rains have been reported to be responsible for fast spread of the bacterium through splashes.

Management

1. Rotate the field with non-cruciferous crops for at least two years.
2. Use disease free seed.
3. Treat the seed with hot water (52° C) for 30 minutes. The growers may get their seed treated from any Plant Pathology Laboratory or Krishi Vigyan Kendra.

4. In endemic areas, cover field floor either with grass or pine needle mulch.
5. Collect and burn the infected plant debris.
6. Apply sprays Bordeaux mixture and repeat at 10 to 14 days interval.

c) Viral diseases

7. Cauliflower mosaic (cauliflower mosaic virus)

Many strains of the virus induce conspicuous mosaic and mottling of leaves, the severity of which is dependent on the varietal tolerance or susceptibility and environmental conditions. Symptoms are usually much more severe in plants simultaneously infected with Turnip Mosaic *Potyvirus* or Broccoli Necrotic Yellows *Rhabdovirus*.

Symptoms

Vein clearing or banding symptoms are produced on the leaves ultimately leading to mosaic pattern (Plate 7). Initially only young leaves are affected having light green patches on them. Early infection may cause reduction in plant growth leading to stunting.

Virus

The disease is caused by cauliflower mosaic virus. The virus belongs to genus *Caulimovirus* and family *Caulimoviridae*. The isometric capsid has a diameter of 50 nm. The capsid shells of virions are composed of multilayers. Capsid appears round.

Host range

Cauliflower, brussels sprouts, cabbage, garden cress, radish, rapeseed and white mustard are the primary hosts of the virus.

Transmission

Most isolates of the virus are transmitted in a semi-persistent manner by the aphid vectors, *Brevicoryne brassicae* and *Myzus persicae*.

Management

1. Most important control procedure is to limit or prevent virus infection of seedlings in seedbeds by establishing seedbeds in

isolation from susceptible crops or, if this is impractical on small farms, by surrounding the seedbeds with a barrier of cereals (mixture of barley and oats).

2. Early rouging of any infected plants can reduce subsequent spread significantly.
3. Areas where CaMV is endemic and causes severe annual epidemics, the culture of an intervening immune crop can sometimes break the infection cycle.

8. Broccoli necrotic yellows (broccoli necrotic yellows virus)

Symptoms

In cabbages and kale, mild vein clearing and leaf rolling appear at first but the disease later becomes symptomless. The disease is symptomless in broccoli except in co-infection with cauliflower mosaic *caulimovirus*. Brussels sprout also remains symptomless after infection.

Virus

The disease is caused by broccoli necrotic yellows virus. The virus belongs to genus *Cytorhabdovirus* and family *Rhabdoviridae*.

Host range

The natural host range of the virus is restricted. The only host species known to be naturally infected are cabbage, kale, broccoli, cauliflower, Brussels sprout and turnip.

Transmission

Natural transmission occurs by the aphid vector (*Brevicoryne brassicae*).Transmission is persistent and propagation occurs in the insects. Mechanical transmission by sap is also of common occurrence but the virus is not transmitted through seed.

Management

1. Since the virus is persistently transmitted by aphids and propagates in the vector, Management of vector through insecticide sprays will restrict the spread of the virus.

2. Prevention of introduction into new areas with plant material can be achieved by diagnostic assays of plants in quarantine; visual inspection is insufficient.

9. Turnip mosaic (turnip mosaic virus)

Symptoms

The symptoms of the disease develop as chlorotic spots on leaves, mottling followed by systemic vein clearing, mosaic and/or necrosis, leaf distortion and often stunting. Infected plants are often become stunted. Some hosts develop progressive necrosis of leaves, petioles and stem with few virus isolates, leading to plant death, particularly in turnip.

Virus

The disease is caused by turnip mosaic virus. The virus belongs to genus *Potyvirus* and family *Potyviridae.*

Host range

Turnip mosaic virus has a very wide host range infecting cauliflower, Brussels sprout, cabbage, turnip, radish, mallow, rapeseed, broccoli, lettuce and mustard.

Transmission

The virus is transmitted by aphids (*Myzus persicae, Brevicoryne brassicae* and *Aphis gossypii*) in non-persistent styletborne manner. Beetles and various species of Thrips can also spread the virus. There is no latent period and aphids typically remain infective for a short period after acquiring the virus.

Management

Control of turnip mosaic *potyvirus* is quite difficult due to the very wide host range of the virus; the ineffectiveness of insecticides in controlling the spread of non-persistently transmitted viruses, and the lack of immune crop cultivars.

1. Remove infected plant debris as well as eradicate all infected plants from in and around field.
2. Schedule plantings in such a way that peak aphid migration periods are avoided.

3. The isolation of nursery beds where the highly susceptible transplants are raised from TuMV infection sources should be a priority, and is practiced in to ensure virus-free seedlings.
4. Use of mineral oil (1%) or neem formulations (2-3 ml/lt) spray reduces the transmission of virus by aphid vectors.

(B) INSECT-PESTS

1. Cabbage Aphid (*Brevicoryne brassicae*) (Hemiptera: Aphididae)

Identification

These are also called as plant lice. The cabbage aphid always remains in group. The general body colour is bluish green with ashy coating on the body. However, in the freshly moulted nymphs the ashy coating disappears. They may be winged or wingless. Cauda and cornicals are present.

Damage

The colonies of aphids are seen on leaves, buds and inflorescence. Nymphs and adults that suck the cell sap from plant parts cause damage. Due to sucking of cell sap, leaves start twisting and curling of leaves take place. Due to severe attack vitality of plant is reduced which ultimately affects the yield.

Management

1. Install yellow sticky trap @12 traps/ha to monitor winged adults or Moerick yellow water traps to catch winged form.
2. When this aphid appears, a spray of quinalphos 25%EC (0.025%) may be given. The activity of this pest remains low due to low temperature during winter, however, during February, when the temperature rises again a spray of the same insecticides may be given and may be repeated after 15 days if the attack persists.
3. Natural enemies: **Syrphid predators** (*Eupeodes frequens, Betasyrphus serarius, Scaeva pyrastri, Metasyrphus confrater, M. corollae, Episyrphus balteatus, Sphaerophoria indiana*).

4. **Coccinellid predators** (*Coccinella septempunctata, Hippodemia variegata, Menochiles sexmaculata*) and ***Chrysoperla* spp.** are the important predators which help in reducing the population of this aphid.
5. Application of neem formulations @ 2–3 ml/litre of water. Neem seed kernel extract @ 4% also helps in reducing the population of this pest.

2. Cabbage butterfly (*Pieris brassicae*) (Lepidoptera: Pieridae)

Identification

This butterfly has wing expanse of 58 to 62 mm and is pale white on dorsal side. On the apical angle of both the forewings, a black patch is found while on both the hind wings, a black patch appears on the apical angle. In females four black spots (2 on each wing) are found in the centre of the forewing on both dorsal and ventral side but in the male these spots are present only on ventral side.

Damage

Caterpillars cause damage, which in initial instars scrape the leaf surface but in the subsequent instars they eat up entire leaves leaving only the harder midribs from veins. In severe infestation the entire plants are eaten up.

Management

Mechanical method: The yellow egg masses of this pest can be collected and killed by crushing. Since the larvae of this pest feed gregariously in earlier stages so these can be crushed and killed.

Biological control: In nature *Apanteles glomeratus* has been reported as the major mortality factor of this pest. Besides, application of Bt @ 500g/ha is effective in reducing the population of this pest.

Chemical control: Spray of neem seed kernel extract (4%) and application of deltamethrin 2.8%EC (0.0025%) also help in reducing the population of caterpillars.

3. Diamondback moth (*Plutella xylostella*) (Lepidopera: Plutellidae)

Identification

Eggs are very small of pinhead size with light yellow in colour. These are laid in batches of 220 mostly on upper surface of leaves. Full grown larva is about 8-12 mm long and green in colour (Plate-1a). On slight disturbance, the larva shows feign death. Adults are small (about 10 mm in length) grayish brown in colour with three dull pale-white triangular markings on costal margin of each forewing. While sitting, these give the diamond-shape pattern thus called as diamondback moth.

Damage

Larvae cause the damage which in earlier stages (I & II instar) feed in mines on leaves of cabbage/cauliflower whereas in the subsequent instars the larvae feed on the exposed leaves and make holes on leaves and subsequently complete skeletonization of plant (Plate 1b). The larvae even shift to the curds/heads of cauliflower and cabbage resulting in considerable crop loss.

Management

1. Visiting the field every day, looking for any adults or larvae on the plants.
2. Use of sticky traps baited with diamondback moth sex pheromone (12/ha) to trap male adults.
3. Sprinkler irrigation after every third day in evening hours is effective.
4. Planting Indian mustard as trap crop
5. NSKE @ 4.0 per cent and Bt formulation @ 500g/ha are also effective against the diamondback moth.
6. Spray application of chlorantraniliprole 18.5%SC (0.002%) or chlorfenapyr 10%SC (0.015%) or chlorfluazuron 5.4%EC (0.015%) or emamectin benzoate 5%SG (0.002%) or indoxacarb 14.5%SC (0.007%) or flubendiamide 20%WG (0.002%) or flufenoxuron 10%DC (0.008%) or metaflumizone 22%SC (0.3%) or lufenuron 5.4%EC (0.006%) or novaluron10%EC (0.015%) or pyridalyl 10%EC (0.01%) or spinosad 2.5%SC (0.003%).

SYMPTOMS OF VEGETABLE DISEASES

Chapter 2. Nursery Diseases and Insect-Pests

Plate 1a. Pre-emergence damping-off showing patchy appearance

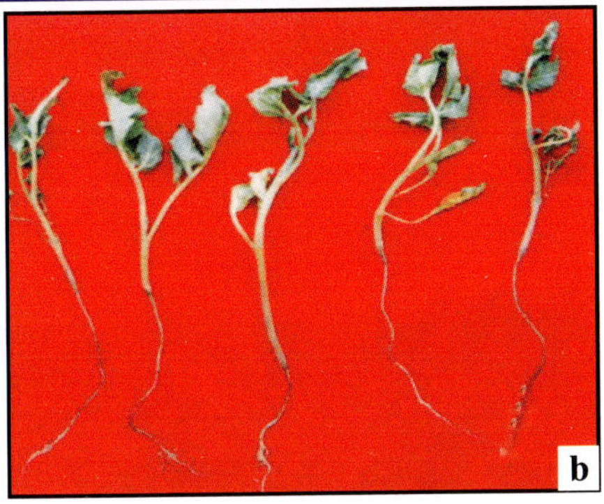

Plate 1b. Post-emergence damping-off

Plate 1

Chapter 3. Diseases and Insect-Pests of Tomato

1a. Infected fruits

1b. Infected fruits fallen on the ground

Plate 1. Symptoms of Buckeye rot of tomato

2a. Infected leaves

2b. Infected fruits

Plate 2. Symptoms of late blight of tomato

3a. Symptoms of *A. solani*

3b. Symptoms of *A. alternata*

3c. Symptoms of *A. a. lycopersici* on leaf

3d. Symptoms of *A. a. lycopersici* on stem

Plate 3. Symptoms of Alternaria leaf spots

Plate 4. Symptoms of Septoria leaf spot

Plate 5. Powdery mildew (*Erysiphe cichoracearum*)

Plate 6. Fusarium wilt (Courtesy N.K. Bharat)

7a. Infected leaf

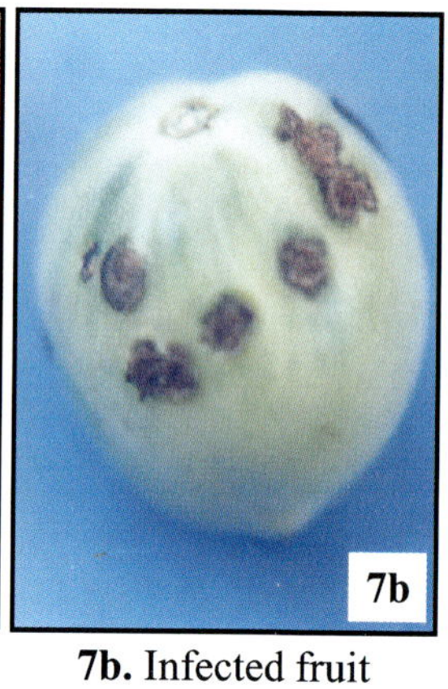

7b. Infected fruit

Plate 7. Symptoms of bacterial spots

Plate 8. Symptoms of bacterial wilt of tomato

9a. On leaves **9b.** On fruit

Plate 9. Symptoms of bacterial canker on tomato

Plate 10. Tomato mosaic

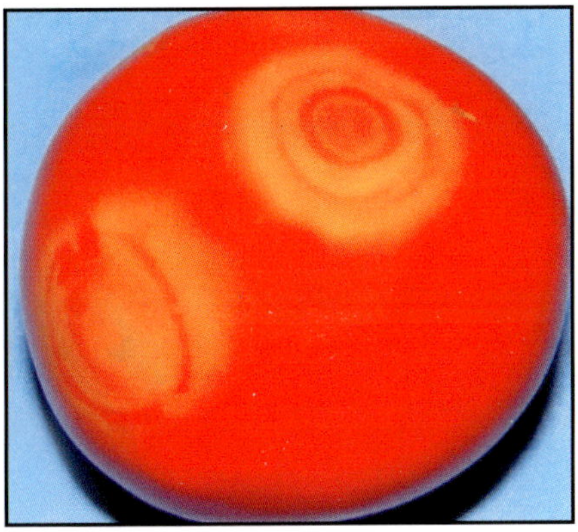

Plate 11. Tomato spotted wilt symptoms on fruit

12a. On Roots **12b.** On Plants

Plate 12. Symptoms of root knot

Chapter 3. Diseases and Insect-Pests of Tomato

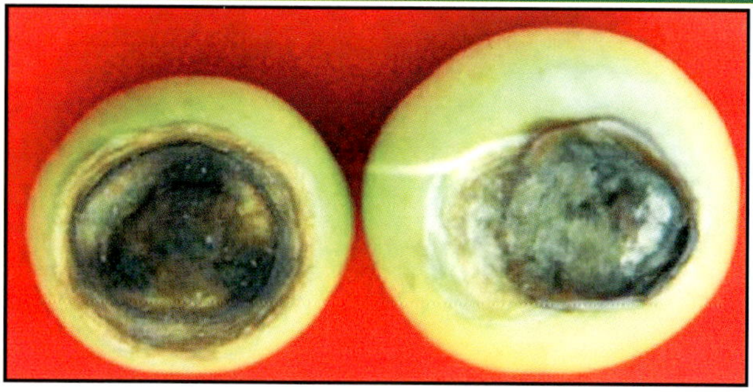

Plate 13. Blossom End Rot

Chapter 4. Diseases and Insect-Pests of Bell Pepper and Chilli

1a. Leaf blight

1b. Fruit rot

Plate 1. Symptoms of leaf blight and fruit rot

Plate 2. Symptoms of Cercospora leaf spot

Plate 3. Symptoms of powdery mildew on the under surface of the leaves

Chapter 4. Diseases and Insect-Pests of Bell Pepper and Chillies

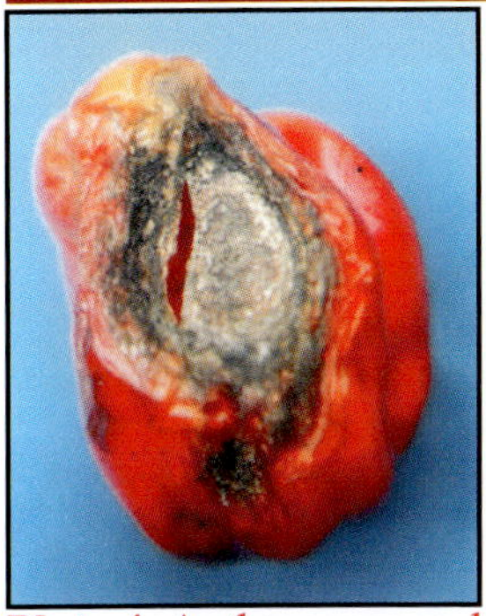

Plate 4. Anthracnose and ripe fruit of bell pepper

(a) On leaves

(b) On fruits

Plate 5. Symptoms of bacterial spot

Plate 6. Symptoms of Chilli Veinal Mottle Virus

Plate 7. Symptoms of Pepper Veinal Mottle Virus

Chapter 5. Diseases and Insect-Pests of Brinjal

1a. On leaves

1b. On fruits

Plate 1. Symptoms of Phomopsis blight on brinjal

Chapter 5. Diseases and Insect-Pests of Brinjal

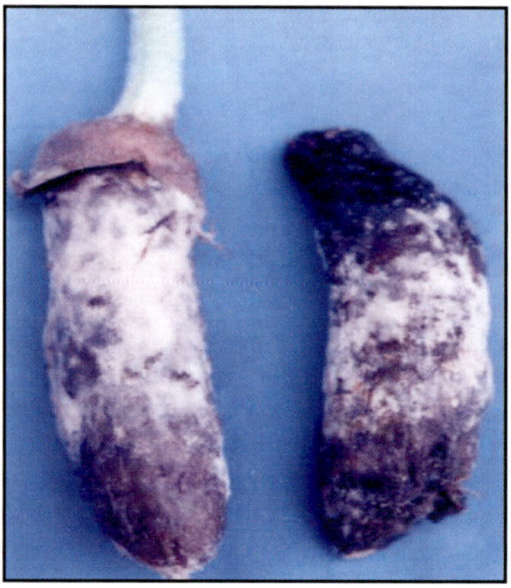

Plate 2. Phytophthora Fruit Rot

Plate 3. Alternaria leaf spot

Chapter 6. Diseases and Insect-Pests of Cucurbits

Plate 1. Symptoms of downy mildew of cucumber

Plate 2. Symptoms of powdery mildew caused by *Sphaerotheca fuliginea*

3b. On Leaves

3b. On fruits

Plate 3. Symptoms of anthracnose

Chapter 6. Diseases and Insect-Pests of Cucurbits

Plate 4. Symptoms of angular leaf spot

Plate 5. Symptoms of cucumber mosaic virus on Cucurbits

Chapter 7. Diseases and Insect-Pests of French Bean

Plate 1. Symptoms of root rot of French bean

2a. On leaves **2b.** On pods

Plate 2. Symptoms of web blight

Plate 3. Symptoms of bean anthracnose

4a. On leaves **4b.** On pods

Plate 4. Symptoms of Angular leaf spot

Plate 5. Symptoms of Floury leaf spot of French bean

Plate 6. Symptoms of bean rust

Plate 7. Symptoms of Crown rot

Plate 8. Symptoms of bean common blight

Plate 9. Symptoms of bean common mosaic

Chapter 8. Diseases and Insect-Pests of Okra

Plate 1. Symptoms of Fusarium wilt of okra

Plate 2. Symptoms of powdery mildew of okra

Plate 3. Symptoms of Cercospora leaf spot on okra

Plate 4. Symptoms of okra yellow vein mosaic

Chapter 9. Diseases and Insect-Pests of Pea

Plate 1. Symptoms of pea powdery mildew

Plate 2. Symptoms of True wilt

Plate 3a. Symptoms of white rot of pea

Plate 3b. Symptoms on pod showing sclerotia

4a. Aecial stage **4b.** Telial stage

Plate 4. Symptoms of pea rust

Plate 5. Symptoms of Ascochyta or *Mycosphaerella* blight of pea

Plate 6. Symptoms of Ascochyta foot rot

Plate 7. Symptoms of Ascochyta pod spot

Chapter 9. Diseases and Insect-Pests of Pea

Plate 8. Symptoms of bacterial blight of pea

Chapter 10. Diseases and Insect-Pests of Crucifers

1a. On Leaf

1b. On Curd

Plate 1. Symptoms of Stalk rot of cauliflower

Plate 2. Curd rot of cauliflower

3a. On leaf

3b. On curd

Plate 3. Symptoms of Downy mildew

Plate 4. Symptoms of Alternaria leaf spot

Plate 5. Symptoms of white rust of crucifers

Plate 6. `V' shape lesions of black rot of crucifers

Plate 7. Symptoms of cauliflower mosaic

Chapter 12. Diseases and Insect-Pests of Onion and Garlic

Plate 1. Symptoms of purple blotch

Plate 2. Stemphylium blight

3a. On bulbs

3b. On leaves

Plate 3. Symptoms of downy mildew

Plate 4. Symptoms of Bulb Rot

1a. On Leaf

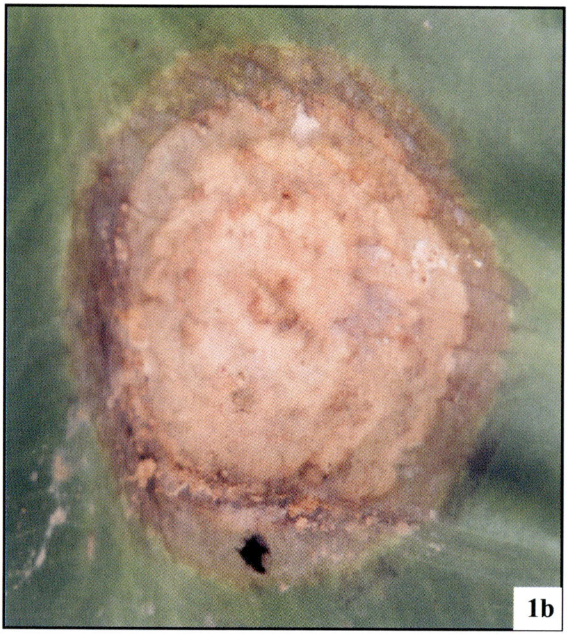

1b. Close up of lesion

Plate 1. Symptoms of Phytophthora blight of Colocasia

DAMAGE OF INSECT-PEST IN VEGETABLE

Chapter 3. Diseases and Insect-Pests of Tomato

Plate 1a. Adult of fruit borer

Plate 1b. Damage of fruit borer on tomato

Plate 2a. Pupae of serpentine leaf miner

Plate 2b. Attack of serpentine leaf miner on tomato

Plate 3. Yellow sticky trap for whitefly

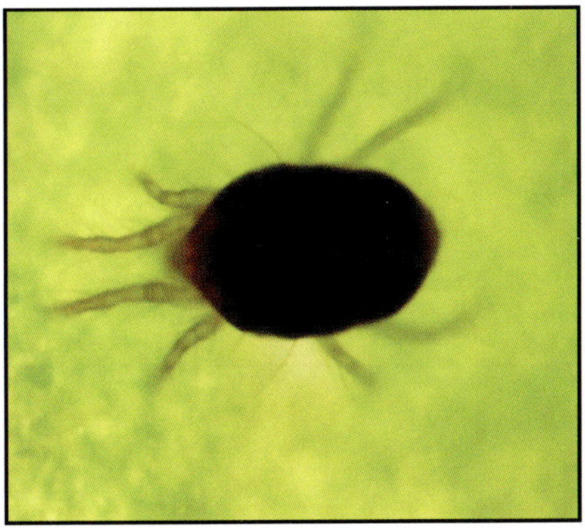

Plate 4. Adult mite

Chapter 4. Diseases and Insect-Pests of Bell Pepper and Chillies

Plate 1. Various stages of *Myzus persicae*

Chapter 7. Diseases and Insect-Pests of French Bean

Plate 1. Damaging symptoms of black bean bug

Chapter 10. Diseases and Insect-Pests of Crucifers

Plate 1a. Larva and pupa of diamondback moth

Plate 1b. Larval infestation of diamondback moth

4. Cabbage Semilooper (*Thysanoplusia orichalcea* and *T. nigrisigna*) (Lepidoptera: Noctuidae)

Identification

The adults of *T. orichalcea* are about 35 mm across the wingspan. The forewings are light brown from centre towards outer margin of wings on dorsal side and with a larger golden patch. The adults of *T. nigrisigna* are about 40 mm across the wingspan. The forewings of this species are dark brown with dirty white patches. The larvae of both these species are stout green with palish tinge and form half loop while walking.

Damage

The damage is done by larvae, which make holes in the leaves. In case of severe infestation, whole plant is defoliated.

Management

1. Use of light traps to attract adults.
2. Mechanical removal of larvae.
3. Application of neem seed kernel extract (4–5%) or neem formulations @ 2–3 ml/litre of water.

5. Crucifer Leaf Webber (*Crocidolomia bionotalis*) (Lepidoptera: Pyralidae)

Identification

Eggs are light green which are laid in groups. Larvae are stout and are light green in colour with whitish longitudinal stripes; three dorsally and one on each lateral side. Adults are light brown in colour with stout body having two white spots on each forewings with fringed margin.

Damage

The larva is the damaging stage which feeds on lower surface of leaves, preferably the young leaves and the growing point. On cabbage head these larvae make hole by making tunnels. The larvae make webbing in severe infestation.

Management

1. Larvae can be killed by manual collection and their subsequent killing or burying deep in the soil.
2. Use of mustard as trap crop.
3. NSKE (Neem Seed Kernel Exract) @ 4.0 per cent or other Neem based insecticides @ 2-3 ml/litre of water.
4. Bt formulations @ 500g /ha.
5. Natural enemy like *Apanteles* sp. takes care of this pest in nature.
6. Spraying on crop with deltamethrin 2.8%EC (0.0025%).

6. Cabbage Head Borer (*Hellula undalis*) (Lepidoptera: Pyralidae)

Identification

Eggs are very small. The larvae are dull white in colour with 7 purplish-brown longitudinal stripes. Adults are light brown with wavy grey markings and a kidney shaped mark on forewing. The hindwings have gray darkening at the margins.

Damage

Damage is caused by larvae which initially mine the leaves and later on feed on the leaf surface sheltered within the silken passage. With the passage, the larvae bore into the heads of cabbage and cauliflower. When the attack is heavy, the plants are riddled with caterpillars

Management

1. Monitoring of pest at early growth stage.
2. Manual collection and destruction of early stage caterpillars.
3. Use of Indian mustard as trap crop.
4. *Bacillus thuringiensis* @ 2g/lit at primordial stage.
5. Application of deltamethrin 2.8%EC (0.0025%).

7. Painted bug (*Bagrada cruciferarum* and *Eurydema pulchrum*) (Hemiptera: Pentatomidae)

Identification

The adults of *B. cruciferarum* have yellow and pink spots on dorsal side of body. Female is comparatively larger in size than male. The length and breadth of adult male is about 5.29 mm and 3.04 mm and in female it is 7.12 mm and 3.94 mm, respectively. The adults of E. pulchrum are dark with red spots on dorsal surface of the body.

Damage

Damage in both the species is done by nymphs as well as adults by sucking sap from the developing pods which gradually wilt and dry up. On leaves, palish or whitish markings appear and such leaves turn brown. When infestation is high, seed formation may be reduced.

Management

1. Follow clean cultivation to avoid infestation since these pests breed on many weeds.
2. The different stages of the bug usually congregate on pods, leaves and stem thus these parts can be shaken vigorously to dislodge them and can be killed in kerosenised water.
3. The remains of the plants should be burnt to kill the different stages of the bug.
4. This pest can be controlled by spraying of imidacloprid 17.8%SL (0.01%).

8. Mustard Sawfly (*Athalia lugens proxima*) (Hymenoptera: Tenthredinidae)

Identification

The adults of this insect can be identified having yellow spot on pronotum, whereas all the four wings are light brown in colour. The larvae are light green in colour with well-developed prominent head. The larvae have three thoracic legs and 6-7 pairs of abdominal legs. They feed gregariously in the earlier instar but feed individually in the later instar.

Damage

Larva is the damaging stage, however, adults also cause damage by making slit on stem by its sharp ovipositor. The damage is caused by defoliating the leaves from the margins inwards. The larvae are generally active during dawn and dusk.

Management

1. Hand picking of larvae and their destruction.
2. Irrigation of field results in drowning of the larvae.
3. Use of neem seed kernel extract @ 5% or neem formulations @ 2–3 ml/lt of water.
4. *Bt* formulations give excellent control.
5. Spray application of quinalphos 25% EC (0.025%) is effective in controlling this pest.

Disclaimer

Though utmost care was taken in recommending only approved Pesticides for management of different diseases and insect-pests, pesticide recommended wrongly if any, may be purely unintended, inadvertent and by oversight. Further no pesticide can be used in commercial scale without CIBRC approved label claim on the target crop.

11

DISEASES AND INSECT-PESTS OF CARROT

Among members of family Umbelliferae, carrot (*Daucus carota* L.) is an important root vegetable and grown throughout the world. Besides temperate countries, carrots are also grown in the tropical countries like India at high altitudes during summers while in plains during winters. This crop is attacked by various diseases and insect-pests, which reduce the potential yields drastically. The present chapter deals with the commonly occurring diseases and insect -pests along with their management methods.

(A) DISEASES

a) Fungal diseases

1. Alternaria blight (*Alternaria dauci*)

Symptoms

Symptoms of the disease first appear on the foliage as small, dark-brown to black spots along the leaf margin and tend to be more elongated, irregular in shape and lack definite margin. Moist weather is favourable for disease development and its spread. It also causes damping-off of seedlings, black decay of roots and blight of seed stalks.

Pathogen

The disease is caused by the fungus *Alternaria dauci*. The pathogen(s) belongs to kingdom Fungi, phylum Ascomycota, class Deutromycetes, order Moniliales and family Dematiaceae.The conidiophores are short and dark. Conidia are dark coloured and have multiseptate character typical of *Alternaria* sp. The broader end of the conidium is attached to the conidiophores.

Predisposing factors

The fungus is both seed and soil borne. It requires free moisture in the form of prolonged dew or rain alongwith 27.8° C temperature for infection.

Management

1. Collect and destruct the infected plant debris.
2. Use healthy seed produced in disease free areas.
3. Seed treatment with hot water (50° C) for 20 to 30 minutes.
4. Spray the crop with Borcdeaux Mixture 4:4:50. Repeat at 10 to 14 days interval.

2. Stemphyllium root rot (*Stemphyllium radicinum*)

Symptoms

Lesions on the roots are circular to irregular, shallow and slightly depressed, with sharp margins and have greenish black to black mealy tissues. Decay at the crown extends deep in to the core. Fungus attacks older leaves. Symptoms produced on leaves and petioles are similar to those of *A. dauci* but the lesions are more widely distributed.

Pathogen

The disease is caused by the fungus *Stemphylium radicinum*. The pathogen belongs to kingdom Fungi, phylum Deuteromycota, class Hypomycetes, order Moniliales and family Moniliaceae. Dark blackishbrown to black colonies contain conidiopho res which arise singly or in groups. Conidiophores are simple and unbranched with one to several conidial scars. Conidia are solitary or bore in chains of two or three; they are of variable shape from elliptical to pear-shaped, with the broadest end attached to the conidiophores. The conidia have 3 to 7 transverse septa and one to several longitudinal septa, possibly with a constriction at each septum.

Predisposing factors

The fungus overwinters in infected seed and in soil as plant debris. Optimum temperature for carrot infection is 27.7° C.

Management

1. During harvesting of roots, avoid injury.
2. Use healthy seed.
3. Drenching of roots with Bordeaux mixture (4:4:50) reduce root rot.

3. Cercospora leaf spot (*Cercospora carotae*)

Symptoms

On the leaf, sub circular, grey to brown or black spots occur at first on the lobes. As the spots increase in number, the intervening parts die and the foliage shrivels. The spots on the petioles and stems are elliptic with a pale centre and dark margin. The fungus also attacks floral parts and late infections permit the fungus to enter the seed.

Pathogen

The disease is caused by *Cercospora carotae*. The pathogen belongs to kingdom Fungi, phylum Ascomycota, class Deutromycetes, order Miniliales and family Dematiaceae.The mycelium of the fungus accumulates in sub-stomatal spaces in the leaf as interwining mass of hyphae. From these stromatic structures, fascilcles of conidiophores develop and emerge through the leaf stomata. The conidiophores are indeterminate in growth and show scars where conidia were attached. The conidia are slightly obclavate, hyaline and several celled.

Predisposing factors

The fungus perpetuates in the infected seed as well as plant debris. Infection by conidia takes place through stomata at optimum temperature (28° C).

Management

1. Field sanitation and long crop rotations help in reducing the disease.
2. Use healthy seed.
3. Spray the crop with Bordeaux mixture (4:4:50) and repeat at 10 to 14 days interval.

4. Powdery mildew (*Erysiphe heraclei*)

Symptoms

Younger and inner leaves are spotted first with white weft like mats of mycelium that soon coalesce forming larger areas and finally covering the entire leaf as it matured. Slight necrosis or no necrosis occurs when plants are slightly affected. When young plants are attacked they are either killed or their vigour is greatly reduced.

Pathogen

The disease is caused by *Erysiphe heraclei.* The pathogen belongs to kingdom Fungi, phylum Ascomycota, class Ascomycetes, order Erysiphales and Family Erysiphaceae. Fungal growth appears as typical white, dense, persistent powdery mildew colonies on leaves, inflorescence, and stems. Conidiophores are long with straight foot cells. Conidia develop either singly or in chains, are cylindrical to ovate. Germ tubes are simple and formed from the ends of conidia. The cleistothecia are initially globular in shape but then become flattened, produced in a scattered or thinly grouped manner. Appendages are narrow to rather coarse, mycelium like and are often interlaced with the mycelium. In each cleistothecium 3-10 asci develop. Each ascus contains 3-6 ascospores.

Predisposing factors

In areas where cleistothecia of the fungus are formed, this explains the mode of perpetuation. Otherwise the fungus perpetuates in the form of conidia from one season to other either on perennial weeds or on other related crops grown in the area. Seed transmission has also been reported.

Management

1. Collect and destruct the affected plant debris.
2. Use resistant cultivars, if available.
3. Spray the crop with fungicides like wettable sulphur (0.2%) and repeat at 10-14 days interval.

(b) Bacterial diseases

5. Soft Rot (*Erwinia carotovora* sub.sp. *carotovora*)

Symptoms

On the lower part of roots, the disease appears as small, water- soaked, brown spots. Later, these spots develop rapidly and spread into the core of the root from where they progress from inside to outside. Lateron, infected part of the root becomes soft due to maceration of tissues, which emit foul smell.

Pathogen

The disease is caused by *Erwinia carotovora* sub. sp. *carotovora*. It is a weak parasite and facultative anaerobe. The bacterium is Gram negative rod, with peritrichous flagella. It produces grayish white colonies on nutrient agar medium and cloudy growth in nutrient broth. It produces gas in the nutrient medium utilizing several carbohydrates as carbon source.

Predisposing factors

Bacteria survive in soil on infected plant debris and from there may infect the growing crop especially after heavy rains or in badly drained fields. Roots may be contaminated any time during harvesting or storage.

Management

1. Improve drainage of the field.
2. Wash harvested roots with chlorinated water to avoid contamination of healthy roots.
3. Dry root surface properly before storage because the bacterium is sensitive to desiccation.
4. Store roots at temperatures close to 0° C.

(B) INSECT–PESTS

Aphids: In general, all aphids have cauda and cornicles

1. Willow carrot aphid: *Cavariella aegopodii* (Hemiptera: Aphididae)

Identification

These are small, soft bodied insects having green or yellowish green in colour. The tips of antennae and apices of legs are brownish in colour.These are generally found in groups.

2. Green peach aphid: *Myzus persicae* (Hemiptera: Aphididae)

Identification

Dealt under the Insect Pests of Bell Pepper and Chili

Damage

Both nymphs and adults cause damage by piercing their mouth parts on plant parts and then by sucking the sap. Due to continuous sucking of sap, vitality of plant is reduced. Besides, some aphids also act as vectors of plant diseases.

Management

Dealt under the Insect Pests of Bell Pepper and Chili.

3. Aster leafhopper: (*Macrosteles quadrilineatus*) (Cicadellidae Hemiptera)

Identification

Adults are very small, wedge shaped and have six spots on head. These are olive green in colour and are very active.

Damage

Damage is caused by both nymphs and adults by piercing their mouth parts due to which the leaves are whitened and give mottled appearance and ultimately disfiguring of leaves takes place. This insect transmits plant disease like yellow phytoplasma which causes aster yellows disease.

Management

1. Planting of susceptible crops near the main crops should be avoided.
2. Destruction of weeds and habitat management.

4. Carrot weevil (*Listronotus oregonensis*) (Curculionidae: Coleoptera)

Identification

Adults are dark brown with copper colour scales on the body. The adults show feign death. Larvae are C–shaped and are white to pinkish white.

Damage

Damage is caused by both adults and grubs, however the destruction of plants is more due to larvae (grubs). The adults feed on leaves and petioles while the larvae make tunnel through roots of plants thus completely damage the plant.

Management

1. Crop rotation may reduce the population of this pest.
2. Since the carrot weevil overwinters in crop residues, thus tilling the residual crop material and removing the weedy margin is effective in managing this pest.

5. Carrot rust fly (*Psila rosae*) (Psilidae Diptera)

Identification

These are small fly of about 5 mm long with reddish brown head.

Damage

The damage is done by the larvae which feed on carrot root due to which the roots start rotting which ultimately leads to secondary infection.

Mangement

Use of row cover to protect the plant from egg laying.

Disclaimer

Though utmost care was taken in recommending only approved pesticides for management of different diseases and insect-pests, pesticide recommended wrongly if any, may be purely unintended, inadvertent and by oversight. Further no pesticide can be used in commercial scale without CIBRC approved label claim on the target crop.

◆

12

DISEASES AND INSECT- PESTS OF ONION AND GARLIC

Onion (*Allium cepa*) and garlic (*Allium sativum*) are two important spice crops which are grown throughout the country and are important foreign exchange earner for our country. During cultivation these crops are infected by several diseases and insect-pests which cause substantial yield losses. Different diseases and insect-pests infecting these crops and their management strategies are described here.

(A) DISEASES

a) Fungal diseases

1. Purple Blotch (*Alternaria porri*)

Symptoms

On leaves and seed stalks, the symptoms of the disease appear as small water-soaked lesions which quickly develop white centers. Later on these lesions become zonated and dark brown to purple in colour (Plate 1). At the point of infection, mostly seed stalks break down. During moist weather, lesion surface may be covered with black sporulation of the pathogen.

Pathogen

The disease is caused by *Alternaia porri*. The pathogen belongs to kingdom Fungi, phylum Ascomycota, class Deutromycetes, order Moniliales and family Dematiaceae. Sexual stage of the fungus is unknown. Conidiophores arise singly or in groups and are pale to mild brown in colour. Conidia are muriform and are usually solitary. Each conidium has 8–12 transverse and zero to several longitudinal or oblique septa. Existence of chlamydospores in the fungus is also reported.

Predisposing factors

The pathogen perpetuates in infected plant debris as dormant mycelium. Rain or persistent dew is important for sporulation and infection. For disease development 25° C temperature is optimum.

Management

1. Fields with disease history should be rotated with unrelated crops.
2. Destroy the infected plant debris.
3. Treat the seed with captan (0.3%).
4. With the initiation of the disease spray the crop with tebuconazole (0.6%) or difenconazole (0.05%) alone or in combination with azoxystrobin (0.1%) or metiram + pyraclostrobin (0.3%) and repeat at 10 days interval.

2. Stemphylium Blight (*Stemphyllium vesicarium*)

This disease is important both in onion and garlic and cause substantial yield losses if the infection takes place in early stage of plant growth.

Symptoms

In onion, the disease symptoms first appear on the radical leaves as small, yellow or pale orange flecks or streaks in the middle of leaf which soon become elongated, spindle shaped to ovate spots, surrounded by characteristic pinkish margin. These spots turn grey at the centre and then become brown to dark olive brown with the development of conidiophores and conidia of the pathogen. Similar symptoms appear on the inflorescence stalks. In garlic, Initial symptoms of the disease appear in the month of January and February as purplish green coloured elongated spots either on the leaves/ stalks resulting in blighting (Plate 2).

Pathogen

The disease is caused by *Stemphylium vesicarium*. The pathogen belongs to kingdom Fungi, phylum Deuteromycota, class Hypomycetes, order Moniliales and family Moniliaceae. Conidiophores are straight to variously curved, 1-4 septate with frequent nodular swellings. The conidia of the fungus are oblong or broadly oval, sometimes inequilateral and are light or medium golden brown to olive brown in colour, having six transverse septa and 1-3 longitudinal septa. Perithecia mature within 3-6 months. Asci are cylindrical to clavate shape. Young ascospores are ellipsoidal and upper half is narrowly tapered.

Predisposing factors

The pathogen perpetuates in the infected plant debris. The disease initiation and further development is favoured by moderate temperatures and high humidity coupled with slight rainfall.

Management

1. Collect and destroy the infected plant debris.
2. Spray the crop with tebuconazole (0.06%) or the mixture of azoxystrobin + difenconazole (0.3%) and repeat at fortnightly interval. Initiate sprays in the first week of February.

3. Downy Mildew (*Peronospora destructor*)

Symptoms

Oval to cylindrical spots of paler than normal green colour are formed on the leaves. The spots may contain alternating green and chlorotic zones (Plage 3b). During moist weather, white to purplish downy growth develops on the under surface of these spots (Plate 3a).

Pathogen

The disease is caused by *Peronospora destructor.* The pathogen belongs to kingdom Chromista, phylum Oomycota, class Oomycetes, order Peronosporales and family Peronosporaceae. The sporangiophores are aseptate, hyaline, swollen at the base and dichotomously branched 3-4 times. Sterigmata are acute to subacute which bear pyriform to fusiform sporangia. Oospores develop late in the season.

Predisposing factors

The fungus perpetuate in the diseased bulbs used for propagation in the next year and oospores present in the diseased crop residues. The fungus requires cool moist nights and only moderately warm days for best development.

The optimum temperature and relative humidity for conidial germination and infection is 25° C and 99–100 per cent, respectively.

Management

1. Collect and burn the infected plant debris.
2. Use healthy seed/ planting material for propagation.
3. With the initiation of the disease spray the crop with azoxystrobin + difenconazole (0.3%) followed by sprays of Zineb 75WP (0.25%) or Bordeaux mixture (4:4:50). Repeat sprays at 10– 14 days interval.

4. Bulb rot (*Fusarium oxysporum* f.sp. *cepae* and *F. solani*)

Symptoms

Characteristic symptoms are rotting of the bulbs, leaf drying from the tip and the affected plants also show wilting symptoms (Plate 4). When cut, the affected tissues appear brown and watery.

Pathogen

The fungi responsible for this disease are *Fusarium oxysporum* f.sp. *cepae* and *F.solani*. The pathogen(s) belongs to kingdom Fungi, phylum Ascomycota, class Deutromycetes, order Moniliales and family Tuberculariaceae. These fungi produce microconidia, macroconidia and chlamydospores.

Predisposing factors

Both *Fusarium* species responsible for this disease are good soil inhabitants and are disseminated by infected onion bulbs and garlic cloves. The disease is prevalent in those areas which have high soil temperature (25–28° C) and moisture.

Management

1. Follow long crop rotation with non–host crops.
2. Dip the seedlings in the solution of zineb 75WP (0.25%) before transplanting.
3. Spray- drench the crop before harvesting with Zineb 75WP (0.25%).

(b) Viral diseases

5. Onion yellow dwarf (onion yellow dwarf virus)

Symptoms

The most characteristic symptoms evoked by the virus are mosaic of leaves, especially mature leaves, and stunting of the whole plant. Leaf striping and curling may also be observed.

Virus

The disease is caused by onion yellow dwarf virus. The virus belongs to genus *Potyvirus* and family *Potyviridae.* The virus is filamentous flexuous with clear modal length with a particle length of 823 nm.

Host range

Primary hosts of the virus are onion, garlic and leek. Ornamental species of *Allium* are also susceptible to this virus. Strains of the virus isolated from each species affected are not transmitted to other species within the genus *Allium.*

Transmission

In nature, the virus is transmitted in a non-persistent manner by aphid *Myzus persicae.* The virus is maintained by vegetative propagation of infected plants. OYDV is experimentally transmissible by mechanical inoculation. The virus is neither transmitted by contact between plants, nor by seed, or pollen.

Management

There are no means of managing this disease other than the use of meristem tip culture for the production of virus-free plants.

6. Leek yellow stripe (leek yellow stripe virus)

Symptoms

Due to systemic infection, generally all plant parts become affected soon after infection, which may result from virus introduction by aphids, as in garlic plants obtained virus-free after meristem-tip culture. Plants derived from vegetative propagation, as all plants of garlic are subject to systemic infection from the onset of growth. In onion, irregular yellow striping of the whole leaf lamina takes place, but mostly at its base. Entire leaves of the plant may become evenly yellow. Affected plants become stunted than normal. Infected plants are less juicy (with their leaves slightly flaccid), the white lower plant part is luster less, and the keeping quality of the harvested product from such plants is very low. Diseased plants are more susceptible to frost and may be killed prematurely.

In garlic, the chlorotic to yellow striping or streaking of leaves is similar to that in onions. These symptoms differ considerably in severity according to virus isolate and cultivar infected. Plant growth may be significantly reduced and yield severely impaired.

Virus

The disease is caused by leek yellow stripe virus. The virus belongs to genus *Potyvirus* and family *Potyviridae.*

Host range

In nature, leek yellow stripe virus is almost limited to leek. Onions and shallots are highly resistant, but not immune to the virus. The garlic-specific strain of the virus, alone or in combination with the garlic-specific strain of onion yellow dwarf *potyvirus,* often occurs in complex and is a major cause of garlic mosaic, usually affecting all garlic plants wherever grown.

Transmission

Transmission through contact and by plant handling is unlikely because the leaves are covered with wax and experimental transmission of virus in sap is often difficult. The virus is transmissible through aphids (*Aphis fabae* and *Myzus persicae*) in the non-persistent manner, but is undoubtedly spread in nature by many migrant aphid species that do not settle on *Allium* species but briefly probe on them.

Management

1. Crops grown from seed or from vegetative planting material should be grown in isolation from infested crops of the same species.
2. The removal of diseased plants may reduce secondary spread within the crop.

3. Aphid control with insecticides is generally found ineffective against non-persistently aphid-transmitted viruses, and spraying with oil is too expensive and not permitted for crops grown for human consumption.

7. Lettuce Necrotic Yellows (lettuce necrotic yellows virus)

Symptoms

Leaves of the infected plants first become dull green and then chlorotic. The outer leaves become flaccid, bronzed and often necrotic. Systemic yellowing and leaf malformation can be observed. Plants become stunted, often severely if infected in the early seedling stage. Late infected plants develop internal necrosis. The virus also causes inter-veinal chlorosis and slight stunting.

Virus

The disease is ccaused by lettuce necrotic yellows virus. The virus belongs to genus *Cytorabdovirus* and family *Rhabdoviridae*.

Host range

Lettuce necrotic yellows virus has a narrow host range. The virus causes serious disease only in lettuce, its main crop host. However, natural infection also occurs in garlic and chickpea. Lettuce necrotic yellows virus is widespread in the weed sowthistle, its principal host, which remains symptomless.

Transmission

The virus is circulative and propagative and is persistently transmitted. The aphid *Hyperomyzus carduellinus* has been shown to colonize sowthistle and transmit the virus. However, not all aphids in a colony are capable of virus transmission.

Management

1. Use of reflective mulches and alternative crop interplanting to alter the apparent crop colour.
2. The alteration of lettuce planting times to avoid peak aphid migration periods has limited applicability due to weather requirements for optimal growth of the crop.
3. Rogueing of weeds and application of insecticides close to crops is also an option.

(B) INSECT-PESTS

1. Onion Thrips (*Thrips tabaci*) (Thysanoptera: Thripidae)

Identification

These are very small of about 1 mm in length. The body is slender with yellowish brown in colour. Wings are trophied or reduced in male while these are well developed in female and have fringe of hairs. The abdomen tapers posteriorly.

Damage

The damage is done by both adults and nymphs (which have asymmetrical types of mouth parts) by lacerating the leaf tissue and then by feeding on the plant juice which comes out by lacerating it. Presence of this insect can be seen by the brisk movement of adults and nymphs on leaves. Attacked leaves show silvery white sheen due to which the leaf curls and dries up. Severely infested plants fail to form flower and bulb. Discolouration and drying up of the leaf tip is the prominent symptom of this pest.

Management

1. Install sky blue colour sticky traps@ 25/ha.
2. Predators like *Orius albidipenis, Chrysoperla* sp. *and* predatory thrips also help in reducing the population of this pest.
3. Use neem coated urea to reduce the infestation of the pest.
4. Spray of dimethoate 30% EC (0.03%) or fipronil 80% WG (0.015%) or lambda cyhalothrin 5% EC (0.005%). A waiting period of 7-10 days should be observed before harvest.

2. Onion Maggot (*Delia anique*) (Diptera: Anthomyiidae)

Identification

Adults are small flies (app. 5mm long), greyish brown in colour. Larvae are cream or dull white in colour.

Damage

It is generally known as root maggot. Damage is more in moist soils. The larvae bore into the bulb which cause yellowing of leaves.

The boring leads to secondary infection by bacteria due to which rotting of bulb takes place.

Management

1. Field sanitation
2. Rotation of crops in the field or field locations may be helpful in reducing onion maggot populations.
3. Deep ploughing of field to expose pupae since this pest pupates in soil.
4. Apply chlorphyriphos 20%EC (0.08%) in soil.

Disclaimer

Though utmost care was taken in recommending only approved pesticides for management of different diseases and insect-pests, pesticide recommended wrongly if any, may be purely unintended, inadvertent and by oversight. Further no pesticide can be used in commercial scale without CIBRC approved label claim on the target crop.

13

DISEASES AND INSECT-PESTS OF COLOCASIA

Colocasia (*Colocasia esculenta* L.Schott), a member of family Araceae, commonly known as taro, is grown throughout the world including India either as staple or subsistence crop. The whole crop is not only used as vegetable but also has medicinal and industrial values. It is rich in proteins, minerals, fibre, phosphorus, iron etc. The starch, mucilage and tissues of this crop possess good industrial value both in textile, paper and alcohol industries. Like other crops, this crop also suffers from the attack of large number of diseases and insect-pests both in field as well as in storage. A comprehensive account of diseases and insect-pests including their management has been given in this chapter.

(A) DISEASES

a) Fungal diseases

1. Phytophthora Blight (*Phytophthora colocasiae*)

Symptoms

The symptoms of the disease appear on leaves as small, dark brown, round spots, which increase in size by centrifugal growth and become circular or assume an oval or irregular shape (Plate 1a). Clear, yellow liquid drops ooze out from the surface of the spots. The periphery of the spots has zonations of different shades of brown, green and yellow and during moist conditions may contain a delicate white haze of sporangial growth (Plate 1b). Badly infected leaves fall down pre-maturely on the soil surface. The corms are also infected by the sporangia which are washed down by rain from the fallen leaves and begin to rot. Sometimes, the corms may rot completely during the unusually wet weather in the field itself but the rot develops more rapidly later on in the storage.

Pathogen

The fungus responsible for this disease is *Phytophthora colocasiae*. The pathogen belongs to kingdom Chromista, phylum Oomycota, class Oomycetes, order Peronosporales and family Pythiaceae. The mycelium is hyaline, coenocytic, intercellular in freshly attacked tissues and intracellular when rotting has advanced. The haustoria are slender, long and unbranched. Sporangiophores are formed on the surface of the leaf and are unbranched, very slender and extremely narrow at the tip. They bear single pear-shaped or elongated sporangia, which germinate directly or indirectly depending on the weather conditions. When they germinate indirectly as many as 12 reniform, bi-flagellate zoospores are formed in each sporangium. Chalamydospores are also formed. These are thick walled, round and hyaline structures. The oogonium is spherical and yellowish in colour. The amphigynous antheridium persists at the base of the oogonium that measures 99 µm in diameter. The oospores are spherical, 20-28 µm in diameter and lie free in the oogonium.

Predisposing factors

The pathogen perpetuates through the mycelium or oospores either present in the crop debris in soil or in infected corms which are used for planting. The disease is severe when wet weather coupled with moderately warm temperature is prevalent.

Management

1. Follow 2-3 years crop rotation.
2. Use healthy seed corms for sowing.
3. Field sanitation and destruction of badly affected plants.
4. Improve drainage of the field.
5. Spray the crop with Bordeaux mixture (4:4:50).

2. Fusarium rot (*Fusarium* spp.)

Many species of *Fusarium* are reported to be associated with this disease and cause losses both in field as well as in storage.

Symptoms

F. oxysporum causes a whitish grey spongy soft rot with a

distinct brown margin. Decay usually begins at the base or side of the corms that may be partially or completely destroyed.

F. solani causes a dry corky or powdery rot with a definite margin. White mould may also develop on the surface and the diseased corm becomes lighter in weight than the healthy one.

Pathogen

Several species of *Fusarium* like *F. oxysporum, F. solani* and *F. proliferatum* are responsible for this disease. The pathogen belongs to kingdom Fungi, phylum Ascomycota, class Deutromycetes, order Moniliales and Family Tuberculariaceae. The fungus produce microconidia, macroconidia and chlamydospores.

Predisposing factors

Fusarium spp. are soil inhabitants and perpetuate in the form of thick walled resting spores (Chlamydospores). *Fusarium* spp. produce abundant asexual spores (conidia), which can cause infection of the corms usually through wounds, sustained during harvesting. Warm temperature (30° C) and humid (90-100% RH) conditions of storage and continuous darkness favour conidial germination.

Management

1. Do not plant Colocasia in the same field year after year.
2. Use healthy seed corms.
3. Treat seed corms in hot water (50° C) for 5 minutes.
4. Select only healthy corms for storage.

(B) INSECT- PESTS

Aphids are abundantly found to infest this plant. Management of aphid is described in other chapters of this Masnual.

Disclaimer

Though utmost care was taken in recommending only approved pesticides for management of different diseases and insect-pests, pesticide recommended wrongly if any, may be purely unintended, inadvertent and by oversight. Further no pesticide can be used in commercial scale without CIBRC approved label claim on the target crop.

✦

14

FUNGICIDES IN VEGETABLE DISEASE MANAGEMENT

Fungicides play an important role in keeping the diseases of vegetables under check in the present day intensive agricultural system and are instrumental in boosting production level of the crops by minimizing the attack of dangerous plant pathogens. Word 'fungicide' is made up of two Latin words i.e. fungus and caedo. The meaning of caedo is "to kill". Thus literally speaking a fungicide would be any agency, which has the ability to kill a fungus. According to this meaning, physical agencies like heat and ultraviolet light should also be called as fungicides. However, that is not so. By common usage, the word is restricted to chemicals. Hence the word fungicide should mean a chemical capable of killing fungi.

Most of the fungicides used before 1970s were multisite control compounds of low biochemical specificity that were limited in their action to surface protection. Fungicides of this type, which were not highly selective against pathogens, included inorganic sulfur and copper compounds such as dithiocarbamates, captan group, organomercurials, organotins, chloronitrobenzens, chlorothalonil, dichlone etc. Use of these compounds depended upon their exclusion from the living plant protoplast in order to control disease without damaging the host. These are the so called protective fungicides that must be applied in anticipation of infection but have little or no therapeutic effect on disease development after infection. During the past three decades, the research and use of agricultural fungicides has been undergoing a dynamic development. The need for increased potency and for internal therapeutic action to eradicate established infections led to the development of biochemically more specific fungicides usually characterized by a narrower antifungal spectrum and susceptibility to fungal resistance problems. These are called as modern fungicides belonging to different chemical classes which are highly selective and mostly systemic in nature having more distinct spectra of activity compared to earlier multisite inhibitors.

COMMONLY USED FUNGICIDE FORMULATIONS

Several fungicides are used for controlling plant diseases. Different fungicides are used in various ways like seed or soil treatment, foliar sprays and post-harvest treatment. It is important that a particular fungicide is used for the specific purpose for which it is effective. Fungicides are broadly available in different forms that can be broadly grouped as follows:

1. **Dusts**

   These formulations contain 4–10 per cent active ingredient and applied in dry form as dusts. They should be light enough to be carried by a slight breeze for a considerable but controlled distance. They should reach and stick to foliage and must be free flowing so that frequent physical agitation is not needed.

2. **Granules**

   These are formulations of toxicants with inert materials formed particles about the size of coarse sugar. The percentage of active ingredient is usually low. Granules have the advantage that they can be measured in dry form more easily than dusts or wettable powders.

3. **Wettable powders (WP)**

   The fungicides in this form are finely powdered. After mixing the fungicide in water they do not settle at the bottom and do not require continuous agitation or pumping during the application time.

4. **Emulsifiable concentrates**

   These are liquids in which the active ingredient is dissolved in a solvent.The fungicides and solvent will often not mix with water, so an emulsifying material is included. When these are added to water a milky mixture is formed which is a suspension of the active ingredient and the emulsified solvent in water.

On the basis of their use in the management of vegetable diseases, commonly used fungicides can be broadly classified as:

1. Seed treatment fungicides.
2. Soil treatment fungicides.
3. Foliar spray fungicides.

1. Seed treatment fungicides

Common Name	Trade Name(s), Dose	Remarks
Captan	Captaf, Hexacap (3g/kg seed)	Mostly used for seed treatment. Also used in the management of the diseases caused by *Pythium, Rhizoctonia* and other soil fungi in vegetable nurseries.
Thiram	Hexathir, Thiride, Arasan, TMTD (2.5-3 g/kg seed)	Mostly used for seed treatment. Also used in the management of the diseases caused by *Pythium, Rhizoctonia* and other soil fungi in vegetable nurseries.
Carboxin	Vitavax (2g/kg seed)	Mostly used as seed dresser for the management of seed borne diseases caused by *Rhizoctonia.*
Carbendazim	Bavistin, Derosal, Bengard (2g/kg seed)	Used for the management of internally seed borne pathogens like *Rhizoctonia, Ascochyta, Fusarium* etc.
Tebuconazole	Raxil (1-2g/kg seed)	Used as seed dresser for the management of smuts, bunts, *Septoria* spp. etc.
Difenoconazole	Dividend (1 -2 g/kg seed)	Used as seed dresser for the management of smuts, bunts, *Septoria* spp. etc.

2. Soil treatment chemicals/ fungicides

Common Name	Trade Name(s), Dose	Remarks
Formaldehyde	Formalin (5% or 1 part Formaldehyde: 7 parts water)	Applied for disinfection of nursery bed soil
Captan	Captaf, Hexacap (0.3%)	Applied as drench for the management of soil borne nursery diseases

3. Foliar spray fungicides

Common Name	Trade Name(s), Dose	Remarks
Non-systemic fungicides		
Sulphur dust	Sulphur dust	Useful against powdery mildew diseases
Wettable Sulphur	Sulfex, Thiovit (0.2%)	Useful against Powdery mildew diseases
Bordeaux mixture	Copper sulphate : Lime : Water (4:4:50)	Useful against Phytophthora blights, downy mildews, leaf spots and bacterial diseases
Copper oxychloride	Blitox, Fytolan (0.3%)	- do-
Cuprous oxide	Cuprous oxide (0.3%)	- do-
Ziram	Cuman L (0.25-0.3%)	Used as protective spray against diseases of vegetables
Zineb	Dithane Z-78, Indofil Z- 78 (0.25)	Commonly used against leaf spots, blights.
Mancozeb	Dithane M-45, Indofil M-45 (0.25%)	Useful for the management of foliar diseases of vegetables

Common Name	Trade Name(s), Dose	Remarks
Propineb	Antracol (0.25-0.3%)	Commonly used against foliar diseases like leaf spots and blights
Chlorothalonil	Kavach (0.2%)	Useful against blights and leaf spots
Iprodione	Rovral (0.25%)	Useful against diseases caused by *Alternaria* spp.
Dinocap	Karathane (0.06%)	Useful against powdery mildew diseases
Systemic Fungicides		
Carbendazim	Bavistin, Derosal, Bengard, Agrozim (0.05- 0.1%)	Useful against leaf spots, powdery mildews, *Rhizoctonia* and *Fusarium* diseases.
Thiophanate methyl	Topsin-M, Roko (0.05-0.1%)	- do -
Metalaxyl + mancozeb	Ridomil MZ (0.25%)	Useful against diseases caused by *Phytophthora*, *Pythium* and downy mildews
Cymoxanil + mancozeb	Curzate M-8 (0.25%)	- do -
Fenarimol	Rubigan (0.04%)	Useful against powdery mildews, rusts and leaf spots
Triadimefon	Bayleton (0.05-0.1%)	- do -
Bitertanol	Baycor (0.05%)	- do -
Propiconazole	Tilt (0.05%)	- do -
Flusilazole	Punch (0.04%)	- do -

Common Name	Trade Name(s), Dose	Remarks
Difenconazole	Score (0.03%)	- do -
Myclobutanil	Systhane (0.05%)	- do -
Hexaconazole	Contaf (0.05%)	- do -
Penconazole	Topas (0.05%)	- d0 -
Tebuconazole	Folicur (0.06%)	- do -
Azoxystrobin	Amistar	Useful against powdery mildews, rusts, downy mildews, Phytophthora diseases.

Precaution: Don't apply systemic fungicide alone twice. Either rotate the fungicide of other group or use in combination of multisite fungicide to avoid the risk of resistance development.

CALCULATIONS FOR PREPARATION OF PESTICIDAL SOLUTIONS IN EC FOR FIELD APPLICATIONS

Emulsifiable Concentrates (EC)

Suppose we want to prepare a solution of an insecticide (say Ethion 50 EC) having field strength of 0.05%. The quantity of insecticide (ethion) to be mixed in 1 litre of water will be determined as follow:

$$\text{Quantity of insecticide (in ml)} = \frac{\text{Desired Concentration (\%)}}{\text{\% ai in EC (given)}} \times \text{Voloume of water (in ml)}$$

$$\text{Quantity of ethion required} = \frac{0.05}{50} \times 1000$$

$$= 1 \text{ ml.}$$

So mix 1 ml. of Ethion (50 EC) in 1000 ml. (1 litre) of water.

15

LIST OF PESTICIDES WHICH ARE BANNED, REFUSED REGISTRATION AND RESTRICTED IN USE

GOVERNMENT OF INDIA MINISTRY OF AGRICULTURE & FARMERS WELFAREDEPARTMENT OF AGRICULTURE, COOPERATION & FARMERS WELFARE DIRECTORATE OF PLANT PROTECTION, QUARANTINE & STORAGE (http: //ppqs.gov.in/divisions/cib-rc/registered-products)

Insecticides / Pesticides Registered under section 9(3) of the Insecticides Act, 1968 for use in the Country:

(As on 31/10/2019)

S.No.	Name of the Pesticide
1.	2,4-D Amine salt
2.	2,4-Dichlorophenoxy Acetic Acid
3.	Abamectin*
4.	Acephate
5.	Acetamiprid
6.	Afidopyropen
7.	Alachlor {To be phase out 31st December, 2020 and no new certificate of registration shall be issued vide S.O. 3951(E) dated 8th August, 2018}
8.	Allethrin
9.	Alphacypermethrin
10.	Alphanaphthyl Acetic Acid
11.	Aluminium Phosphide
12.	Ametroctradin

S.No.	Name of the Pesticide
13.	Ametryn
14.	*Ampelomycesquisqualis*
15.	Anilophos
16.	Atrazine
17.	Aureofungin
18.	Azadirachtin (Neem Products)
19.	Azimsulfuron
20.	Azoxystrobin
21.	*Bacillus sphaericus*
22.	*Bacillus subtillus*
23.	*Bacillus thuringiensis var.* galleriae
24.	*Bacillus thuringiensis var.* israelensis
25.	*Bacillus thuringiensis var.* kurstaki
26.	Barium Carbonate
27.	Beauverlabassiana
28.	Benalaxyl (TIM)
29.	Benalaxyl M
30.	Bendiocarb
31.	Benfuracarb
32.	Bensulfuron Methyl
33.	Bentazone TI
34.	Beta Cyfluthrin
35.	Bifenazate
36.	Bifenthrin
37.	Bispyribac Sodium
38.	Bitertanol
39.	Boscalid+ Pyraclostrobin
40.	Bromadiolone
41.	Buprimate (FI-WRT)
42.	Buprofezin
43.	Butachlor

S.No.	Name of the Pesticide
44.	Captan
45.	Carbendazim
46.	Carbofuran
47.	Carbosulfan
48.	Carboxin
49.	Carfentrazone Ethyl
50.	Carpropamid
51.	Cartap Hydrochloride
52.	Chlorantraniliprole
53.	Chlorfenopyr
54.	Chlorfluazuron
55.	Chlorimuron ethyl
56.	Chlormequat Chloride (CCC)
57.	Chlorothalonil
58.	Chlorpropham (TI),TIM
59.	Chlorpyriphos
60.	Chlorpyriphos Methyl
61.	Chlothianidin
62.	Chromafenozide
63.	Cinmethylene
64.	Clodinafop-propargyl
65.	Clodinafop-propargyl+Sodiumacifluorfen
66.	Clomazone
67.	Clothianidin
68.	Copper Hydroxide
69.	Copper Oxychloride
70.	Copper Sulphate
71.	Coumachlor
72.	Coumatetralyl
73.	Cuprous Oxide
74.	Cyantraniliprole

S.No.	Name of the Pesticide
75.	Cyazofamid
76.	Cyenopyrafen (FI–WRT)
77.	Cyflumetofen
78.	Cyfluthrin
79.	Cyhalofop-butyl
80.	Cymoxanil
81.	Cypermethrin
82.	Cyphenothrin
83.	Cyproconazole (TI)
84.	Dazomet
85.	Deltamethrin (Decamethrin)
86.	Diafenthiuron
87.	Dichloro Diphenyl Trichloroethane (DDT)
88.	Dichloropropene and Dichloropropanemixure (DD mixure)
89.	Dichlorvos (DDVP) {To be phase out 31st December, 2020 and no new certificate of registration shall be issued vide S.O. 3951(E) dated 8th August, 2018}
90.	Diclofop-Methyl
91.	Diclosulam
92.	Dicofol
93.	Difenoconazole
94.	Diflubenzuron
95.	Dimethoate
96.	Dimethomorph
97.	Dinocap
98.	Dinotefuron
99.	Dithianon
100.	Diuron
101.	Dodine
102.	D-trans Allethrin
103.	Edifenphos

S.No.	Name of the Pesticide
104.	Emamectin Benzoate
105.	Epoxyconazole
106.	Ethephon
107.	Ethion
108.	Ethiprole
109.	Ethofenprox (Etofenprox)
110.	Ethoxysulfuron
111.	Ethylene Dichloride and Carbon Tetrachloride mixture (EDCT Mixture 3: 1)
112.	Etoxazole(FI)
113.	Famoxadone
114.	Fenamidone
115.	Fenazaquin
116.	Fenitrothion
117.	Fenobucarb (BPMC)
118.	Fenoxaprop-p-ethyl
119.	Fenpropathrin
120.	Fenpyroximate
121.	Fenvalerate
122.	Fipronil
123.	Flonicamid
124.	Fluazifop-p-butyl
125.	Flubendiamide
126.	Flucetosulfuron
127.	Fluchloralin
128.	Fluensulfone 47% TC (MUP) (FI)
129.	Flufenacet
130.	Flufenoxuron
131.	Flufenzine
132.	**Flumioxazin**
133.	Fluopicolide

S.No.	Name of the Pesticide
134.	Fluopyram and its metabolite
135.	Flupyradifurone*
136.	Flusilazole (TI)
137.	Fluthiacet methyl (TIM)
138.	Fluvalinate
139.	Fluxapyroxad 167 g/L + Pyraclostrobin 333g/l SC (FI -WRT)
140.	Fomesafen
141.	Forchlorfenuron
142.	Fosetyl-Al
143.	Gibberellic Acid
144.	Glufosinate Ammonium
145.	Glyphosate
146.	Haloxyfop-R-methyl 10.55%.EC(FI)
147.	Helosulfuron methyl
148.	Hexaconazole
149.	Hexazinone
150.	Hexythiazox
151.	Hydrogen Cyanamide
152.	Imazamox
153.	Imazethapyr
154.	Imidacloprid
155.	Imiprothrin
156.	Indoxacarb
157.	Indaziflam+ Glyphosate ammonium (FI)
158.	Iprobenfos (Kitazin)
159.	Iprodione
160.	Iprovalicarb
161.	Isoprothiolane
162.	Isoproturon
163.	Kasugamycin
164.	Kresoxim Methyl

S.No.	Name of the Pesticide
165.	Lambdacyhalothrin
166.	Lime Sulphur
167.	Lufenuron
168.	Magnesium Phosphide Plates
169.	Malathion
170.	Mancozeb
171.	Mandipropamid
172.	Mepiquate Chloride
173.	Meptyldiinocop
174.	Mesosulfuron Methyl + Iodosulfuron Methyl Sodium
175.	Metaflumizone
176.	Metalaxyl
177.	Metalaxyl-M
178.	Metaldehyde
179.	Metamifop TI
180.	Metamitron (TIM)
181.	*Metarhiziumanisopliae*
182.	Methabenzthiazuron
183.	Methomyl (12.5% L& 24% formulationsare banned **for import, manufacture and use**)
184.	Methoxyfenazide (FI- WRT)
185.	Methyl Bromide
186.	Methyl Chlorophenoxy Acetic Acid (MCPA)
187.	Metiram
188.	Metofluthrin
189.	Metolachlor
190.	Metrafenone
191.	Metribuzin
192.	Metsulfuron Methyl
193.	Milbemectin
194.	Monocrotophos

S.No.	Name of the Pesticide
195.	Myclobutanil
196.	Novaluron
197.	Nuclear polyhyderosis virus of *Helicoverpaarmigera*
198.	Nuclear polyhyderosis virus of *Spodopteralitura*
199.	Orthosulfamuron
200.	Oxadiargyl
201.	Oxadiazon
202.	Oxathiapipron
203.	Oxycarboxin
204.	Oxydemeton-Methyl
205.	Oxyfluorfen
206.	Paclobutrazol
207.	Paraquat dichloride
208.	Penoxasulam
209.	Penconazole
210.	Pencycuron
211.	Pendimethalin
212.	Penflufen*
213.	Penoxsulam
214.	Permethrin
215.	Phenthoate
216.	Phorate {To be phase out 31st December, 2020 and no new certificate of registration shall be issued *vide* S.O. 3951(E) dated 8th August, 2018}
217.	Phosalone
218.	Phosphamidon {To be phase out 31st December, 2020 and no new certificate of registration shall be issued vide S.O. 3951(E) dated 8th August, 2018}
219.	Picoxystrobin'TIM
220.	Pinoxaden
221.	Prallethrin
222.	Pretilachlor

S.No.	Name of the Pesticide
223.	Primiphos-methyl
224.	Prochloraz TI
225.	Profenophos
226.	Prohexadione Calcium
227.	**Propamocarb hydrochloride technical 66% w/w min (Aqueous concentrate)**
228.	Propanil
229.	Propaquizafop
230.	Propergite
231.	Propetamphos
232.	Propiconazole
233.	Propineb
234.	Propoxur
235.	*Pseudomonas fluorescens*
236.	Pymetrozin (FI), TIM
237.	Pyraclostrobin
238.	Pyraclostrobin+Epoxiconazole
239.	Pyrazosulfuron ethyl
240.	Pyrethrin (pyrethrum)
241.	Pyridaben (FI- WRT)
242.	Pyridalyl
243.	Pyriproxyfen (TI)
244.	Pyrithiobac sodium
245.	Pyroxasulfon(FI- WRT)
246.	Quinalphos
247.	Quizalofop ethyl
248.	Quizalofop-P-tefuryl
249.	S-bioallethrin
250.	Sodium paranitrophinolate
251.	Spinetoram
252.	Spirotetramat

S.No.	Name of the Pesticide
253.	Spinosad
254.	Spiromesifen
255.	Streptomycin + Tetracycline
256.	Sulfentrazone TIM
257.	Sulfosulfuron
258.	Sulfoxaflor
259.	sulphur
260.	Tebuconazole
261.	Tembotrione
262.	Temephos
263.	Tetraconazole (FI)
264.	Thiacloprid
265.	Thifluzamide
266.	Thiobencarb (Benthiocarb)
267.	Thiocyclam Hydrogen oxalate
268.	Thiodicarb
269.	Thiomethoxain
270.	Thiophanate-Methyl
271.	Thiram
272.	Tolfenpyrad (TIM)
273.	Topramezone
274.	Transfluthrin
275.	Triacontanol
276.	Triadimefon
277.	Triafamone (Triafamone 20% w/w + Ethoxysulfuron 10% WG % w/w SC FI)
278.	Triallate
279.	Triasulfuron
280.	Triazophos {To be phase out 31st December, 2020 and no new certificate of registration shall be issued vide S.O. 3951(E) dated 8th August, 2018}

S.No.	Name of the Pesticide
281.	Trichlorfon {To be phase out 31st December, 2020 and no new certificate of registration shall be issued vide S.O. 3951(E) dated 8th August, 2018}
282.	*Trichodermaharzianum*
283.	*TrichodermaViride*
284.	Tricyclazole
285.	Triflumezopyrim (TIM)
286.	Trifloxistrobin
287.	Trifluralin {The registration, import, manufacture, formulation, transport, sell and its alluses except use in wheat shall prohibitedand completely banneedvide S.O. 3951(E) dated 8th August, 2018}
288.	Validamycin
289.	*Verticilliumlecanii*
290.	Zinc Phosphide
291.	Zineb
292.	Ziram

1. PESTICIDES AND FORMULATIONS REGISTERED FOR USE IN THE COUNTRY UNDER THE INSECTIICDES ACT, 1968:

(Updated on 31.10.2019)

S.No.	Name of the Pesticides	Formulation registered	No. of formu-lation
1.	2 ,4-Dichlorophenoxy Acetic Acid (2,4-D Sodium Amine and Ester Salt	a) 2, 4-D Sodium Salt used as Tech a.i. 80% w/w min.	1
		b) 2,4-D Amine Salt 58% SL 22.5% SL	2
		c) 2,4-D Ethyl Ester 38% EC, 4.5% Gr., 20%WP,	3
2.	Abamectin	1.9%EC, 1.9%SC	2
3.	Acephate	75% SP, 95%SG, 97.0% DF	3

S.No.	Name of the Pesticides	Formulation registered	No. of formu-lation
4.	Acetamiprid	20 SP	1
5.	Afidopyropen	50 g/l DC (FI-WRT)	1
6.	Alachlor {To be phase out 31st December, 2020 and no new certificate of registration shall be issued *vide* S.O. 3951(E) dated 8th August, 2018}	50%EC, 10% GR	2
7.	Allethrin	0.5% Coil, 4% Mat, 0.5% Aer., 3.6% L, 0.2% & 0.02% Coil	6
8.	Alphacypermethrin	10% EC, 5% WP, 0.5% Chalk, 10% SC, 0.1% RTU, 0.55% LLIBN, 0.667% LLIN, 0.66% W/W LLIN	8
9.	Alphanaphthyl Acetic Acid	4.5% Sol.	1
10.	Aluminium phosphide *(R)	56% Tab, 56% Powder,77.5% Gr., 6% Tab	4
11.	Ampliomycesquisqualis	2.00%WP	1
12.	Anilofos	30% EC, 18% EC	2
13.	Ametryn	80%WG (FI)	1
14.	Atrazine	50% WP	1
15.	Aureofungin	46.15% SP	1
16.	Azadirachtin (neem products)	25%, 10%, 0.03% EC, 0.1% EC, 0.15% EC, 5% EC, 0.3%EC, 15% extract concentrâtes, 1% EC, 0.1% Gr, 0.15%Gr, 0.03% (300 PPM) w/w min	12
17.	Azimsulfuron	50% DF (F.I.)	1

S.No.	Name of the Pesticides	Formulation registered	No. of formu-lation
18.	Azoxystrobin	23% SC (FI), **23% SC (FIM)**	2
19.	Bacillussubtillis	2.0% AS	1
20.	Bacillus thuringiensis var. sphaericus	1.3% FC	1
21.	Bacillus thuringiensis var. galleriae	1.3% FC	1
22.	Bacillus thuringiensis var. israelensis	Liquid & WP formulations, 5% AS, 12% AS, 5.0% WP	3
23.	Bacillus thuringiensis var. kurstaki	5% WP, 2.5% As, 0.5% WP, 3.5% AS,	4
24.	Barium Carbonate	1% P	1
25.	Beauveriabassiana	1.15% WP, 1.0% WP, 1.15%SC, 10% SC, 5.0% WP	5
26.	Bendiocarb	80% WP	1
27.	Benfuracarb	40%EC, 3.0% GR	2
28.	Beta cyfluthrin	2.45% SC	1
29.	Bensulfuron Methyl	60% DF(FI)	1
30.	Bentazone	480 g/l SL	1
31.	Bifenazate	50% WP(FI) 22.6% SC(FI)	2
32.	Bifenthrin	10% EC, 2.5%EC, 23.4%, MUP (Imp), 8% SC(FI), 0.05% MC (11 Hrs.), 10% P,0.05%MC (8 hrs), 8% SC (FIM)	8
33.	Bispyribac sodium	10% SC (FI), 10% w/v SC (FIM)	2
34.	Bitertanol	25% WP	1
35.	Bromadiolone	0.25% CB, 0.005% RB & 0.005% RB cake	3
36.	Buprimate	26.7% w/w EC FI	1
37.	Buprofezin	25% SC, 70 .0 % DF	2

S.No.	Name of the Pesticides	Formulation registered	No. of formu-lation
38.	Butachlor	50% EC, 5% Gr., 50% EW,	3
39.	Captan	50% WP, 75% WP, 50% WDG	3
40.	Carbendazim	25% DS, 50% WP, 46.27% SC, 5% Gr	4
41.	Carbofuran	3% CG,	1
42.	Carbosulfan	25% DS, 25% EC, 6% Gr.	3
43.	Carboxin	75% WP	1
44.	Carfentrazone-ethyl	40% EC, 40% DF(FI), 53%MUP(FI)	3
45.	Carpropamid	27.8% SC	1
46.	Cartap Hydrochloride	4% Gr., 50% SP, 75%SG	3
47.	Chlopropham	50% HN	1
48.	Chlorantraniliprole	18.5% SC, 0.4% Gr.	2
49.	Chlorfenapyr	10% SC (FI)	1
50.	Chlorfenuron	10%SC(FI)	1
51.	Chlorfluazuron	5% EC w/w, 5% w/v, 5.4% EC(FI) & FIM	3
52.	Chlorimuron ethyl	25% WP,	1
53.	Chlormequat Chloride	50% Sol.	1
54.	Chlorothalonil	75% WP	1
55.	Chlorpyriphos	20% EC, 10% Gr,., 1.5% DP, 50% EC, 2% RTU, 20.0% w/w CS	6
56.	Chlorpyriphos Methyl	40% EC	1
57.	Chromafenozide	80% WP	1
58.	Cinmethylene	10% EC	1
59.	Clodinafop-propargyl (Piroxofop-propinyl)	15% WP	1
60.	Clomazone	50% EC	1
61.	Clothianidin	50%WG(FI), 50% WG formulation	2

S.No.	Name of the Pesticides	Formulation registered	No. of formulation
62.	Copper Hydroxide	77% WP, 53.8% DF(FI)	2
63.	Copper Oxychloride	50% WP, 40% Paste, 5% DP, 50 WG, 56% OP	5
64.	Copper Sulphate	Used as Tech 98% equivalent to cooper content 25% w/w min., 2.62% SC	2
65.	Coumachlor	0.5% CB, 0.025% RB	2
66.	Coumatetralyl	0.75% TP, 0.037% Bait.	2
67.	Cuprous Oxide	4% DP	1
68.	Cyantraniliprole	10.26% OD	1
69.	Cyazofamid	34.5% SC (FI)	1
70.	Cyenopyrafen	30.0% w/w SC(FI-WRT)	1
71.	Cyfluthrin	10% WP, 5% EW,	2
72.	Cyflumetofen	20% SC	1
73.	Cyhalofop-butyl	10% EC	1
74.	Cymoxonil	80% WP, 50% WP	2
75.	Cypermethrin	10% EC, 25% EC, 1% Chalk, 0.1% Aquous (HH), 0.25 DP, (3% Smoke Generator-to be used only through Pest Control Operators and not allowed to be used	6
76.	Cyphenothrin	5% EC, 0.15% in combination as Aer. , 7.2% VP	3
77.	Dazomet	Dazomet Technical (soil Sterilant GR) not permitted on tea	1

S.No.	Name of the Pesticides	Formulation registered	No. of formu-lation
78.	Decamethrin (Deltamethrin)	2.5% Flow2.5% WP2.8% EC, 0.5% Chalk, 1.25% ULV, 25% Tab., 11% EC, 0.5% Tablet, 1.8% EC, **2%"w/w EW**	10
79.	Diafenthiuron	50% WP, 47.8% SC	2
80.	Dichloro Diphenyl Trichloroethane (DDT)*R	50% WP, 75% WP	2
81.	Dichloropropene and Dichloropropanes mixture (DD Mixture) *(R)	1: 1	1
82.	Diclofop-methyl	28% EC	1
83.	Diclosulam	84% WDG(FI)	1
84.	Diclorvos (DDVP) {To be phase out 31st December, 2020 and no new certificate of registration shall be issued vide S.O. 3951(E) dated 8th August, 2018}	76% EC	1
85.	Dicofol	18.5% EC	1
86.	Difenoconazole	25% EC, 3% WS	2
87.	Diflubenzuron	25% WP,2%Tab (FI) 2% Tab(FIM),2%Gr,(FI), 2% Gr(FIM)	5
88.	Dimethoate	30% EC	1
89.	Dimethomorph	50% WP	1
90.	Dinotefuran	20% SG (F.I.)	1
91.	Dinocap	48% EC	1

S.No.	Name of the Pesticides	Formulation registered	No. of formu-lation
92.	Dithianon	75% WP	1
93.	Diuron	80% WP	1
94.	Dodine	65% WP, 50% flow, 40% SC	3
95.	D-trans allethrin	2% Mat, 0.1% coil (8hrs), 0.1% coil (12 hr.), 0.1% Coil(11hrs), 0.1% Coil (6hrs), 21.97% Mos. Mat	6
96.	Edifenphos	50% EC	1
97.	Emamectin Benzoate	5% SG (FI) & (FIM) , 1.9% EC, 0.1% Gel	3
98.	Ethephon	39% SL, 10% Paste	2
99.	Ethion	50% EC	1
100.	Ethofenprox (Etofenprox)	10% EC	1
101.	Ethoxysulfuron	10% EC, 15%WG(FI)	2
102.	Ethylene Dichloride and Carbon Tetrachloride mixture (EDCT mixture 3: 1)	3: 1	1
103.	Etoxazole	10%SC(FI)	1
104.	Fenazaquin	10 EC	1
105.	Fenitrothion*R	2% Spray, 20% OL (**banned** in agriculture use except for locust in scheduled dessert area and public health)	2
106.	Fenobucarb (BPMC)	50% EC	1
107.	Fenoxaprop-p-ethyl	10% EC, 9.3% EC,6.7%EC	3
108.	Fenpropathrin	10% EC, 30% EC	2

S.No.	Name of the Pesticides	Formulation registered	No. of formu-lation
109.	Fenpyroximate	5% EC, 5% SC	2
110.	Fenvalerate	0.4% DP, 20%EC	2
111.	Fipronil	0.3% Gr., 5% SC, 0.05%Gel (Import) & FIM, 80%WG (FI & FIM), 2.92% EC, 0.6% GR, 18.87% w/w SC	7
112.	Flonicamide	50% WG	1
113.	Fluazifop-p-butyl	13.4% EC	1
114.	Flubendiamide	39.35%SC, 20%WG, 0.7% GR	3
115.	Flucetosulfuron	10% WG (FI)	1
116.	Fluchloralin	45% EC	1
117.	Fluensulfone	47% MUP (FI), 2.0% GR,	2
118.	Flufenacet	60% WP	1
119.	Flufenoxuron	10% DC	1
120.	Flumioxazin	50.0% w/w SC	1
121.	Flumite	20% SC	1
122.	Fluopyram	34.48% SC	1
123.	Flupyradiflurone	17.9% w/w SL	1
124.	Flusilazole	40% EC	1
125.	Fluthiacet methyl	10.3% w/w EC (FIM)	1
126.	Fluvalinate	25% EC	1
127.	Fluxapyroxad	333g/l FS (FI)	1
128.	Forchlorfenuron (CPPU)	0.1%, 0.12% EC (FI)	2
129.	Fosetyl-Al	80% WP	1
130.	Gibberellic Acid	Tech. P, 0.186% SP, 0.001% L, 40% WSG	4
131.	Glufosinate Ammonium	13.5% SL, 50.0% TK	2
132.	Glyphosate	41% SL, 20.2% SL, 5%SL	3
133.	Glyphosate ammonium salt	71% SG, 5%SL(FI)	2

S.No.	Name of the Pesticides	Formulation registered	No. of formulation
134.	Glyphosate Isopropyl amine salt (IPA)		1
135.	Halosulfuron methyl	75% WG(FI)	1
136.	Haloxyfop R Methyl	10.5% EC (FI)	1
137.	Hexaconazole	5% EC, 5% SC, 2% SC, 75% WG	4
138.	Hexythiazox	5.45% EC	1
139.	Hydrogen cyanamid	49% age, 50% SL	2
140.	Imazethapyr	10% SL, 70% WG (FI)	2
141.	Imidacloprid	17.8% SL, 70% WS, 48% FS, 30.5% SC, 2.5%Gel, 70%WG, 0.3% Gr, 0.5% RB(FI), 0.03 % w/w Gel(FI)	9
142.	Imiprothrin	50% MUP	1
143.	Indoxacarb	14.5% SC,15.8%EC	2
144.	Iprobenfos (Kitazin)	48% EC, 17% Gr.	2
145.	Iprodione	50% WP	1
146.	Isoprothiolane	40% EC	1
147.	Isoproturon	50% WP, 75% WP, 50% Flow	3
148.	Kasugamycin	3% SL	1
149.	Kresoxim-methyl	44.3%(500 g/l)SC	1
150.	Lambda-cyhalothrin	2.43%CS ,5% EC, 10% WP, 2.5% EC, 0.5% Chalk, 22.8%CS(FI), 4.9%CS, 9.7% CS	8
151.	Lime Sulphur	22% SC	1
152.	Lufenuron	5.4% EC	1
153.	Magnesium phosphide Plates	56%Min. Plates (FI) &FIM	2

S.No.	Name of the Pesticides	Formulation registered	No. of formu-lation
154.	Malathion	5% DP, 25% WP, 50% EC, 0 .25% Spray and 96% ULV, 2% Spray, 5% Spray	7
155.	Mancozeb	75% WP, 35% SC, 75%WG	3
156.	Mandipropamid	23.4% SC	1
157.	Mepiquat Chloride	5%AS, 50% TK	2
158.	Meptyldinocap	35.7% w/w EC	1
159.	Metaflumizone	22% SC (FI)	1
160.	Metalaxyl	35% WS, 40% WS	2
161.	Metalaxyl - M	31.8% ES	1
162.	Metaldehyde	2.5% DP	1
163.	Metamitron	70%SC	1
164.	Metarhiziumanisopliae	1.15 % WP, 1.0% WP, 1.5% Liquid	3
165.	Methabenzthiazuron	70% WP	1
166.	Methomyl (12.5% L & 24% formulations are **banned for import, manufacture and use**)	40% SP	1
167.	Methyl bromide *(R)	99% L, 98% L	2
168.	Methyl chlorophenoxy acetic acid	40% SL or 40% As	1
169.	Methoxyfenazide	21.8% w/w SC (FI)	1
170.	Metiram	70% WG	1
171.	Metofluthrin	0.005% Mosquito coil, 5%EC, 0.32% w/w LV	3
172.	Metolachlor	50% EC	1
173.	Metrafenone	500g/l SC	1
174.	Metribuzin	70% WP	1

S.No.	Name of the Pesticides	Formulation registered	No. of formu-lation
175.	Metsulfuron -methyl	20% WP, 20%WG(FI), 20 % WG FIM	3
176.	Milbemectin	1%EC	1
177.	Monocrotophos*R	36% SL, 15% w/w SG	2
178.	Myclobutanil	15%SG,10%WP	2
179.	Novaluron	10% EC (FI), 8.8% SC, 10% EC	3
180.	NPV of Helicoverpaarmigera	0.43% AS, 2.0% AS. 0.5% AS	3
181.	NPV of Spodopteralitura	0.5% AS	1
182.	Orthosulfamuron	50% WG	1
183.	Oxadiargyl	80% WP, 6% EC	2
184.	Oxadiazon	25% EC	1
185.	Oxathiapipron	10.1% w/w (FI)	1
186.	Oxycarboxin	20% EC	1
187.	Oxydemeton-methyl	25% EC	1
188.	Oxyfluorfen	23.5% EC, 0.35% Gr.	2
189.	Paclobutrazol	23% SC(FI), 23% SG, 40.0% w/w SC	3
190.	Paraquat dichloride	24% SL	1
191.	Penconazole	10% EC	1
192.	Pencycuron	22.9% SC	1
193.	Pendimethalin	30% EC, 5% Gr.,38.7%CS(FI)&FIM	4
194.	Penflufen	22.43% FS,	1
195.	Penoxsulam	21.7% SC(FI), 2.67% OD(FI), 50% w/w MUP, 21.7% SC (FIM v/s FIT),	3

S.No.	Name of the Pesticides	Formulation registered	No. of formu-lation
196.	Permethrin	25% EC, 5% SG., 2% EC, LLIN 2% (FI)	4
197.	Phenthoate	2% DP, 50% EC	2
198.	Phorate {To be phase out 31st December, 2020 and no new certificate of registration shall be issued vide S.O. 3951(E) dated 8th August, 2018}	10% CG	1
199.	Phosalone	4% DP, 35% EC	2
200.	Phosphamidon {To be phase out 31st December, 2020 and no new certificate of registration shall be issued vide S.O. 3951(E) dated 8th August, 2018}	40% SL	1
201.	Picoxystrobin	22.52%SC(FI), 22.5 % w/w SC (FIM v/s FIT)	1
202.	Pinoxaden	5.1% EC	1
203.	Prallethrin	0.8% mat for 12 hours, 1% Mat, 0.8% L, 1.6% L, 0.5% mosquito coil, 0.04% Mosquito coil, 1.2% mat, 19% w/w VP, 0.6% mat, 2.4%LV	10
204.	Pretilachlor	50% EC, 30.7% w/w EC,37.0%EW	3
205.	Primiphos-methyl	25% WP, 50% EC, 1% Spray	3
206.	Profenophos	50% EC	1
207.	Prohexadione calcium	10% WG (FI)	1

S.No.	Name of the Pesticides	Formulation registered	No. of formu-lation
208.	Propanil	35% EC	1
209.	Propaquizafop	10% EC(FI)	1
210.	Propergite	57% EC	1
211.	Propetamphos	20% EC, 1% Spray	2
212.	Propiconazole	25% EC	1
213.	Propineb	70% WP	1
214.	Propoxur	20% EC, 1% Aer., 2% Aer. 1% HH Spray, 2% Bait	5
215.	Pseudomonas flouroscens	0.5%WP,1.75%WP, 1%WP,1.5 %WP, 2.0% AS, 1.5% LF	6
216.	Pymetrozine	50%WG(FI), 50% WG (FIM v/s FIT)	2
217.	Pyrazosulfuron -ethyl	10% WP, 70% WDG(FI)	2
218.	Pyrethrins (Pyrethrum)	0.2% DP, 2.5% EC, 0.05% Spray, 0.2% PH, 2.0% EC	5
219.	Pyridaben	20% w/w WP (FI)	1
220.	Pyridalyl	10% EC	1
221.	Pyriproxifen	0.5% Gr(FI), 0.5% GR (FIM v/s FIT) 10%EC(FI), 10 % EC (FIM)	3
222.	Pyrithiobac Sodium	10% EC	1
223.	Pyraclostrobin	20% WG (FI), 100 g/l CS(FI)	1
224.	Pyroxasulfone	85.0 % w/w WG (FI-WRT)	1
225.	Quinalphos	1.5% DP, 25% EC, 20% AF, 5%Gr	4
226.	Quizalofop ethyl	10% EC, (FI), 5% EC (FI), 15.0% EC (FIM)	3
227.	Quizalofop-p-tefuryl	4.41% EC (FI)	1

S.No.	Name of the Pesticides	Formulation registered	No. of formu-lation
228.	S-Bioallethrin	2.4% mat	1
229.	Sodium paranitrophenolate	1.8% SL(FI), 0.3% SL	2
230.	Spinetoram	11.7% SC (FI)	1
231.	Spinosad	45% SC, 2.5% SC	2
232.	Spiromesifen	22.9% SC	1
233.	Spirotetramat	15.31% OD	1
234.	Streptomycin + Tetracycline	90: 10 SP	1
235.	Sulfentrazone	39.6% SC	1
236.	Sulfoxaflor 21.8% SC(FI)	21.8% SC(FI)	1
237.	Sulfosulfuron	75% WG	1
238.	Sulphur	85% DP, 80% WP, 40% SC, 80% WG/WDG , 55.16 SC (800 gm / L) 40%WP,52%SC	7
239.	Tebuconazole	2.5% DS, 2% DS, 25.9%EC, 25% WG, 5.36% FS, 5.4% (6.0%w/v) FS, 38.39% SC	7
240.	Tembotrione	34.4% SC	1
241.	Temephos	50% EC, 1% Sand Granules	2
242.	Tetraconazole	3.8% EW (FI), 11.6% w/w (12.5% w/v) SL (FI)	2
243.	Thiacloprid	21.7% SC	1
244.	Thifluzamide	24% SC	1
245.	Thiobencarb (Benthiocarb)	50% EC, 10% Gr.	2
246.	Thiocyclam hydrogen oxalate	50% SP, 4% Gr	2
247.	Thiodicarb	75% WP	1

S.No.	Name of the Pesticides	Formulation registered	No. of formu-lation
248.	Thiomethoxam	25% WG, 70% WS, 30%FS, 75% SG, 0.01% Gel bait	5
249.	Thiophanate-methyl	70% WP, 70% w/w WG	2
250.	Thiram	80% WP, 40% FS,75%WS	3
251.	Tolfenpyrad	15%EC	1
252.	Topramizone	336 g/l SC(FI)	1
253.	Transfluthrin	0.88% Liquid Vaporiser, 0.03% Mos. Coil, 20%MV Gel (30 days mat tray), 1% FU, 1.2% LV, 1.6% LV, 12.0 % AE	7
254.	Triallate	50% EC	1
255.	Triasulfuron	20%WG	1
256.	Triazophos {To be phase out 31st December, 2020 and no new certificate of registration shall be issued vide S.O. 3951(E) dated 8th August, 2018}	40% EC, 20% EC, 20.0% WG	3
257.	Trichlorfon {To be phase out 31st December, 2020 and no new certificate of registration shall be issued vide S.O. 3951(E) dated 8th August, 2018}	5% DP, 50% EC, 5% Gr.	3
258.	Trichodermaviride	1% WP (CFU 2x106 gm/min), 0.5% WP, 5%WP	3
259.	Tricodermaharzianum	0.5% WS, 2.0% WP, 1.0% WP	3
260.	Tricontanol	0.05% EC, 0.1% EW, 0.05% GR	3
261.	Tricyclazole	75% WP, 70% WG	2

S.No.	Name of the Pesticides	Formulation registered	No. of formulation
262.	Triflumezopyrim	10% SC	1
263.	Trifluralin {The registration, import, manufacture, formulation, transport, sell and its alluses except use in wheat shall prohibitedand completely banneed vide S.O. 3951(E) dated 8th August, 2018}	48% EC, (All uses of this product shall be completely banned except use in wheat).	1
264.	Validamycin	3% L	1
265.	Verticilliumlecanii	1.15% WP, 1.5% Liquid	2
266.	Verticilliumchlamydosporium	1.00 WP	1
267.	Zinc Phosphide	2% RB, 80% Powder, 1% Bait	3
268.	Zineb	75%WP,80% WP, 27% Colloidal Suspension	3
269.	Ziram	80% WP, 27% SC	2
		Total	**563**

*R: Restricted

2. APPROVED FORMULATION OF COMBINATION PESTICIDES:

S.No.	Combination Product	Company
	A. INSECTICIDES	
1.	Acephate 50% + Bifenthrin 10% WDG	M/s GSP Crop Science Ltd., Ahmedabad
2.	Acephate 45% + Cypermethrin 5% DF	M/s UPL Limited.
3.	Acephate 25% + Fenvalerate 3% EC	M/s Rallis India Ltd., Bangalore
4.	Acephate 50% + Imidacloprid 1.8% SP	M/s United Phosphorus Ltd.

S.No.	Combination Product	Company
5.	Acephate 5% + Imidacloprid 1.1% SP	M/s United Phosphorus Ltd.
6.	Acetamiprid 0.4% + Chlorpyriphos 20% EC	M/s Gharda Chemicals Ltd, Mumbai
7.	Acetamiprid 1.1% + Cypermethrin 5.5% w/w EC	M/s United Phosphorus Ltd.
8.	Beta cyfluthrin 8.49% + Imidacloprid 19.81% OD	M/s Bayer Crop Science Ltd., Mumbai
9.	Bifenthrin 3% + Chlorpyriphos 30% w/w EC	Insecticide India Ltd.
10.	Buprofezin15% + Acephate35% WP	M/s. Makhteshim Agan Pvt. Ltd.
11.	Buprofezin 20% + Acephate 50% w/w WP	M/s Rallis India Ltd., Bangalore
12.	Buprofezin 9.0% + Acephate 24.0% w/w WP	M/s Crystal Crop Protection Pvt. Ltd., Delhi
13.	Buprofezin 22.0% + Fipronil 3.0% SC	M/s Insecticides India Ltd.
14.	Buprofezin 23.1% + Fipronil 3.85% SC	ADAMA India Ltd.
15.	Cartap hydrochloride 50% + Buprofezin 10% WP	M/s Krishi Rasayan Exports Pvt. Ltd.
16.	Chlorantraniliprole 4.3% + Abamectin 1.7% w/w SC	M/s Syngenta India Ltd.
17.	Chlorantraniliprole 8.8% w/w + Thiamethoxam 17.5% w/w SC	M/s Syngenta India Ltd.
18.	Chlorantraniliprole 9.3% + Lambda-cyhalothrin 4.6% ZC	M/s. Syngenta India Ltd.
19.	Chloropyriphos 16% + Alphacypermethrin 1% EC	M/s Acco Industries Ltd. Mumbai
20.	Chlorpyriphos 50% + Cypermethrin 5% EC	M/s De-Nocil, Mumbai

S.No.	Combination Product	Company
21.	Chlorpyriphos 35.0% + Fipronil 3.5% w/w EC	M/s Gharda Chemicals Limited
22.	Cyantraniliprole 19.8% + Thiamethoxam 19.8% FS	M/s. Syngenta India Ltd.
23.	Cyfluthrin 0.025% + Tranfluthrin 0.04%	M/s Bayer India
24.	Cypermethrin 10% +Indoxacarb 10% EC	M/s Gharda Chemicals Limited
25.	Cypermethrin 3% + Quinalphos 20% EC	M/s United Phosphorus Ltd., Mumbai
26.	D-trans-allethrin 0.1% + Permethrin 0.03% +Imiprothroin 0.02% Aerosol	M/s Recktt Benckisar (I) Ltd.
27.	Deltamethrin 0.05% + Allethrin .04% L	
28.	Deltamethrin 0.75% +Endosulfan* 29.75% EC	
29.	Deltamethrin 1% + Triazophos 35% EC	
30.	Difenthiuron 40.1% + Acetamiprid 3.9% w/w	M/s Insecticides India Pvt. Ltd., Delhi
31.	Difenthiuron 47.0% + Bifenthrin 9.4% SC	ADAMA India Ltd
32.	Diflubenzuron 20%+ Deltamethrin 2%SC	M/s. Chemtura Chemicals India Pvt. Ltd., Mumbai
33.	Emamectin benzoate 1.5% + Fipronil 3.5% SC	M/s Crystal crop Protection Pvt. Ltd., Delhi
34.	Emamectin benzoate 3.0% + Thiamethoxam 12.0%WG	M/s UPL Ltd., Mumbai M/ s Insecticides India Ltd.
35.	Endosulfan* 35% + Cypermethrin 5% EC	M/s Excel Industries Ltd., Mumbai
36.	Ethion 40% + Cypermethrin 5% EC	M/s Rallis India Ltd., Bangalore.

S.No.	Combination Product	Company
37.	Ethiprole 40% + Imidacloprid 40% (80% WG)	M/s Bayer Crop Science Ltd, Mumbai
38.	Fipronil 4% + Acetamiprid 4% w/w SC	
39.	Fipronil 7% + Hexythiazox 2% w/w SC	M/s Coromandal international Ltd.
40.	Fipronil 40% + Imidacloprid 40% WG	M/s Bayer Crop Science Ltd., Mumbai
41.	Fenobucarb 20% +Buprofezin 5% SE	M/s Crystal crop Protection Pvt. Ltd., Delhi
42.	Fipronil 4% Thiamethoxam 4% w/w SC	M/s Parijat Industries (India) Pvt. Ltd., N. Delhi
43.	Flubendiamide 4% + Buprofezin20%SC	M/s. Rallis India Ltd. Bangalore
44.	Hexaconazole 4.0% + Carbendazim 16.0% SC	M/s Insecticides India Ltd., New Delhi
45.	Imidacloprid 21% + Beta-cyfluthrin 10.5% SC	M/s Bayer Crop Science Ltd., Mumbai
46.	Imidacloprid 18.5% + Hexaconazole 1.5% FS (Seed Dresser)	M/s Rallis India Ltd., Bangalore
47.	Imidacloprid 6.0% + Lambda cyhalothrin 4.0% SL	M/s Willowood Chemical Pvt. Ltd.
48.	Imiprothrin 0.05% + Cypermethrin 1.0% CL	M/s Godrej Consumer Products Ltd., Mumbai
49.	Imiprothrin 0.07% + Cypermethrin 0.20% Aerosol	M/s Godrej Consumer Products Ltd., Mumbai
50.	Imiprothrin 0.1% + Cyphenothrin 0.15%	
51.	Indoxacarb 14.5% + Acetamiprid7.7% SC	M/s Rallis India Ltd., Bangalore
52.	Isoprothiolane 28% + Fipronil 5% w/w EC	M/s Parijat Industries (India) Pvt. Ltd., N. Delhi

S.No.	Combination Product	Company
53.	Methyl bromide 98% + chlorpicrin 2%	
54.	Novaluron 5.25% + Indoxacarb 4.5% SC	M/s MakhteshimAgan India Pvt Ltd., Hyderabad
55.	Novaluron 5.25% + Emamectin benzoate 0.9% w/w SC	M/s. ADAMA India Pvt. Ltd.,
56.	Phosalone 24% + Cypermethrin 5% EC	M/s Aventis CropscienceLtd..
57.	Phosphamidon 40% + Imidacloprid 2% SP	M/s United Phosphorus Ltd.
58.	Profenofos 40% + Cypermethrin 4% EC	M/s Syngenta India Limited, Pune
59.	Profenofos 40% + Fenpyroximate 2.50%EC	M/s Excel Crop Care Ltd., Mumbai
60.	Propoxur 0.25% + Cyfluthrin 0.025% Aerosal	M/s Bayer India
61.	Propoxur 0.5% + Cyfluthrin 0.015% Spray	
62.	Propoxur 0.5% + Cyfluthrin 0.025% Spray,	
63.	Pyriproxyfen 10% + Bifenthrin 10% w/w SC	M/s Parijat Industries (India) Pvt. Ltd., New Delhi
64.	Pyriproxyfen 10% + Bifenthrin 10% w/w EC	M/s Parijat Industries (India) Pvt. Ltd., N. Delhi
65.	Pyriproxyfen 5.0% + Diafenthiuron 25% SE	M/s GSP crop life Pvt Ltd.
66.	Pyriproxyfen 5% + Fenpropathrin 15% EC	M/s Sumitomo Chemical India Pvt. Ltd.
67.	Spirotetramat 11.01% w/w + Imidacloprid 11.01% w/w SC (FIM)	M/s Bayer CropScience Ltd., Mumbai
68.	Thiamethoxam 0.9%+ Fipronil 0.2% GR	Crystal Crop Protection Pvt Ltd.
69.	Thiamethoxam 12.6% + Lambda-cyhalothrin 9.5% ZC	M/s Syngenta India Ltd.

S.No.	Combination Product	Company
70.	Thiamethoxam 1.0% w/w + Chlorantraniliprole 0.5% w/w GR	M/s Syngenta India Ltd., Pune
	B. FUNGICIDES	
1.	Ametoctradin 27% + Dimethomorph 20.27 %SC	BASF India Ltd., Mumbai
2.	Azoxystrobin 4.8% w/w + Chlorothalonil 40.0% w/w SC	M/s Syngenta India Ltd.
3.	Azoxystrobin 18.2% + Cyproconazole 7.3% w/w SC	M/s Syngenta India Ltd.
4.	Azoxystrobin 8.3%+ Mancozeb 66.7% WG	M/s United Phosphorus Ltd.,
5.	Azoxystrobin11.5%+ Mancozeb 30.0% WG	M/s Crystal Crop Protection Pvt. Ltd., Delhi
6.	Azoxystrobin16.7%+ Tricyclazole 33.3% SC	M/s Crystal Crop Protection Pvt. Ltd., Delhi
7.	Azoxystrobin 11.0%+ Tebuconazole 18.3% SC	M/s GSP Crop Science Pvt. Ltd., Ahmedabad
8.	Azoxystrobin 12.5%+ Tebuconazole 12.5% SC	M/s Excel Crop Care Ltd., Mumbai
9.	Azoxystrobin 7.1 % + Propiconazole 11.9% SE (Import) and FIM	M/s Adama Makhteshim Ltd.
10.	Benalaxyl-M 4.0% + Mancozeb 65.0% WP	Ms Isagro (Asia) Agrochemical Pvt. Ltd.
11.	Benalaxyl 8.0 % + Mancozeb 65% WP	M/s FMC India Ltd.
12.	Boscalid 25.2+ Pyraclostrobin 12.8% WG	M/s BASF India Ltd., Mumbai
13.	Captan 70% + Hexaconazole5%WP	
14.	Carbendazim 12% + Mancozeb 63% WP	M/s United Phosphorus Ltd., Mumbai

S.No.	Combination Product	Company
15.	Carbendazim 12% + Mancozeb 63% WS	M/s United Phosphorus Ltd., Mumbai
16.	Carbendazim 1.92% + Mancozeb 10.08% GR	M/s United Phosphorus Ltd., Mumbai
17.	Carbendazim 25% + Mancozeb 50%WS	M/s Indofil Industries Ltd.
18.	Carbendazim 25% Flusilazole 12.5% SE	M/s DhanukaAgritech Ltd.
19.	Carboxin 17.5% + Thiram 17.5% FF	M/s Chemchura Chemical Pvt. Ltd.
20.	Carboxin 37.5% + Thiram 37.5% WS	M/s Crompton Specialaties Asia Pacific
21.	Carfentrazone Ethyl 20% + Sulfosulfuron 25% WG	M/s FMC India Pvt. Ltd., Bangalore
22.	Cartap hydrochloride 4% + Fipronil 0.5% w/w CG	M/s Willowood Chemicals Pvt. Ltd.
23.	Copper Sulphate47.15 % + Mancozeb30 % WDG	UPL Ltd. Mumbai
24.	Cymoxanil 8%+Mancozeb 64% WP	Dupont Crop Protection Ltd.
25.	Dimethomorph 12% + Pyraclostrobin 6.7% WG (FI)	M/s. BASF India Pvt. Ltd., Mumbai
26.	Famoxadone 16.6% + Cymoxanil 22.1% SC	M/s EI Dupont India Pvt. Ltd., Gurgaon
27.	Fenamidone 10% + Mancozeb 50% WDG(FI)	M/s Bayer Crop Science Ltd.
28.	Fenamidone 4.44% + Fosetyl-Al 66.66% WDG(FI)	M/s Bayer Crop Science Ltd.
29.	Fluocoplide 4.44% + Fosetyl-Al 66.66% WDG(FI)	M/s Bayer Crop Science Ltd.
30.	Fluopicolide 5.56% w/w + Propamocarb hydrochloride 55.6% w/w SC	M/s Bayer Crop Science Ltd.

S.No.	Combination Product	Company
31.	Fluopyram 17.7 % + Tebuconazole 17.7% SC	M/s Bayer CropScience Ltd., Mumbai
32.	Fluxapyroxad 75 g/l + Difenconazole 50g/l SC	M/s BASF India Ltd., Mumbai
33.	Fluxapyroxad 250 g/l + Pyraclostrobin 250 g/l SC(FI)	M/s BASF India Ltd., Mumbai
34.	Fluxapyroxad 167 g/L + Pyraclostrobin 333g/l SC (FI)	M/s BASF India Ltd., Mumbai
35.	fluxapyroxad 62.5 g/l + epoxiconazole 62.5 g/l EC	M/s BASF India Ltd., Mumbai
36.	Hexaconazole 4% + Carbendazim 16% w/w SC	M/s Insecticide India Ltd.
37.	Hexaconazole 5% + Validamycin 2.5% SC	M/s Willowood chem. Pvt Ltd.
38.	Hexaconazole 4% + Zineb 68% WP	M/s Indofil Industries Ltd.
39.	Improvalicarb 5.5% + Propineb 61.25% WP	M/s Bayer Crop Science Ltd.
40.	Iprodion 25% + Carbendazim 25% WP	M/s Aventis Crop Science Ltd., Mumbai
41.	Kasugamycin 5% + Copper Oxychloride 45% WP	M/s. Dhanuka Agritech Ltd.
42.	Kresoxim methyl 15% + chlorothalonil 56% WG	Rallis India Ltd.
43.	Kresoxim methyl 40%+ Hexaconazole w/w WG	Rallis India Ltd.
44.	Mancozeb 40% +Azoxytrobin 7.0% w/w OS	M/s Coromandel International Ltd., Telangana
45.	Mandipropamid 5.0% w/w + Mancozeb 60.0% w/w WG (FI)	M/s Syngenta India Limited

S.No.	Combination Product	Company
46.	Metalaxyl -M 3.3% + Chlorothalonil 33.1% SC	M/s Syngenta India Ltd.,
47.	Metalaxyl -M 8% + Mancozeb 64% WP	M/s Syngenta India Ltd, Pune
48.	Metalaxyl-M 4% + Mancozeb 64% WP	M/s Syngenta India Ltd.
49.	Metiram 55% + Pyraclostrobin 5% WG (FI)	M/s BASF India Ltd.
50.	Metiram 44% + Dimethomorph 9% w/w WG (FI)	M/s BASF India Ltd.
51.	Penflufen l3.28% w/w + Trifloxystrobin 13.28 w/w FS	M/s Bayer Cpyriproamrop Science Ltd.
52.	Picoxystrobin 6.78 %+ Tricyclazole 20.33% SC	M/s E.I. Dupont India Private Ltd.
53.	Picoxystrobin 7.05 % +Propiconazole 11.71%SC	M/s E.I. Dupont India Private Ltd.
54.	Prochloraz 24.4% + Tebuconazole 12.1%w/w EW (FI)	ADAMA India Pvt. Ltd.
55.	Propiconazole 10.7% + Tricyclazole 34.2% SE	M/s Syngenta India Ltd., Pune
56.	Propiconazole 13.9% + Difenoconazole 13.9% EC	M/s Syngenta India Ltd.
57.	Propineb 54.2%+ Tricyclazole 15.0% WP	M/s Coromandel International Ltd., Telangana
58.	Pyroclostrobin 133g/l+ Epoxiconazole 50 g/l (w/v) SE(FI)	M/s BASF India Ltd.
59.	Pyrithiobac Sodium 6% + Quizalofop Ethyl 4% w/w EC	M/s Godrej Agrovet Ltd., Mumbai
60.	Quizalofop ethyl 10% EC + Chlorimuron ethyl 25% WP + Surfactant (0.2) Twin pack	M/s. Dhanuka Agritech Ltd., New Delhi
61.	Quizalofop ethyl 4% + Pyrithiobac sodium 6.0% w/w EC	M/s Godrej Agrovet Ltd., Mumbai

S.No.	Combination Product	Company
62.	Streptomycin + Tetracycline (90+10)	M/s Hindustan Antibiotics Ltd., Mumbai
63.	Tebuconazole 6.7% + Captan 26.9% w/w SC	M/s. ADAMA India Pvt. Ltd.,
64.	Tebuconazole 10% + Sulphur 65% WG	M/s Excel Crop Care Ltd., Mumbai
65.	Tebuconazole 50% + Trifloxystrobin 25% WG (FI)	M/s Bayer Crop Science Ltd., Mumbai
66.	Thiophanate Methyl 450 g/L+Pyroclostrobin 50 g/L FS (FI)	M/s. B ASF India Ltd.
67.	Triafamone 20% w/w + Ethoxysulfuron 10% WG % w/w SC 390 RC	M/s Bayer Crop Science Ltd., Mumbai
68.	Tricyclazole 45% + Hexaconazole 10% WG	M/s Indofil Industries Ltd.
69.	Tricyclazole 18.0 w/w + Tebuconazole 14.4% w/w SC	M/s Dow Agroscience India Pvt. Ltd., Mumbai
70.	Tricyclazole 20.4 %w/w + AZoxystrobin 6.8% SC	M/s Dow Agroscience India Pvt. Ltd., Mumbai
71.	Tricyclazole 18% + Mancozeb 62% WP	M/s Indofil Industries Ltd.
	C. HERBICIDES	
1.	Anilofos 24% + 2, 4-D 32% EC	M/s Aventis Cropscience Ltd., Mumbai
2.	Bensulfuron methyl 0.6% + Pretilachlor 6% Gr	M/s NagarjunaAgrichem
3.	Carfentrazone-ethyl 20% + sulfosulfuron 25% WG	M/s FMC India Pvt. Ltd.,
4.	Carfentrazone ethyl 0.43% + Glyphosate 30.82% w/w EW	M/s FMC India PVt. Ltd., Bangalore
5.	Clodinafoppropargyl 9% + Metribuzin 20% WP	M/s. Crystal Plant Protection P. Ltd.

S.No.	Combination Product	Company
6.	Clodinafop-propargyl 15% + Metsulfuran methyl 1% WP	M/s United Phosphorus Ltd.
7.	Clodinafop-propargyl 16.5% + Sodium acifluorfen 8%WP	M/s United Phosphorus Ltd.
8.	Clomazone 20% + 2,4-diethyl ester 30% EC	M/s FMC India Pvt. Ltd.
9.	Fenoxaprop-p-ethyl 7.77%+ Metribuzin 13.6% EC	M/s Bayer Crop Science Ltd., Mumbai
10.	Fluxapyzoxad 62.5% g/L + Epoxyconazole 62.5% g/L EC	M/s BASF India Ltd., Mumbai
11.	Fomesafen 11.1% w/w + Fluazifop-p-butyl 11.1% w/w SL	M/s Syngenta India Ltd., Pune
12.	Hexazinone 13.2% + Diuron 46.8% WP(FI)	M/s EI Dupont India Pvt. Ltd., Gurgaon
13.	Imazamox 35% + Imazethapyr 35% WG(FI)	M/s BASF India Ltd., Mumbai
14.	Imazethapyr 2%+Pendimethalin 30%EC	M/s. BASF India Ltd., Mumbai
15.	Indaziflam 1.65% + Glyphosate-Isopropyl ammonium 44.63% SC (FI)	M/s Bayer Crop Science Ltd., Thane
16.	Metsulfuron Methyl 10% + carfentrazone ethyl 40% DF	M/s E. I. DuPont India Pvt. Ltd. Gurgaon
17.	Mesosulfuron-methyl 3% + Idosulfuron-methyl sodium 0.6% WG (FI)	M/s Bayer Crop Science Ltd.
18.	Metfulfuron Methyl 10% + Chlorimuron Ethyl 10% WP	M/s E.I. Dupont India Ltd., Gurgaon
19.	Metribuzin 42% + Clodinafoppropargyl 12% + WG	M/s United Phosphorus Ltd.
20.	Oxyfluorfen 2.5% + Isopropyl amine salt of Glyphosate 41% SC	M/s. ADAMA India Pvt. Ltd.,
21.	Penoxsulam 0.97% w/w + Butachlor 38.87% w/w SE	M/s Dow Agroscience India Pvt. Ltd., Mumbai

S.No.	Combination Product	Company
22.	Penoxsulam 1.02% + Cyhalofop butyl 5.1 % OD	M/s Dow Agroscience India Pvt. Ltd., Mumbai
23.	Pretilachlor 6% + Pyrazosulfuron ethyl 0.15% GR	M/s United Phosphorus Ltd.,
24.	Pretilachlor 30.0% + Pyrazosulfuron ethyl 0.75%WG	M/s United Phosphorus Ltd.,
25.	Propaquizafop 5% + Oxyfluorofen 12% EC	M/s. Adama India Pvt. Ltd.,
26.	Propaquizafop 2.5% + Imazethayper 3.75% w/w ME	M/s ADAMA India Pvt. Ltd., Hyderabad
27.	Sodium Aceflourofen 16.5% + Clodinafop-propargyl 8% EC (FI)	M/s United Phosphorus Ltd.
28.	Sulfentrazone 28.0% + Clomazone 30.0% w/w WP	FMC India Pvt. Ltd.
29.	Sulfosulfuron 75% + Metsulfuron 5% WDG	M/s United Phosphorus Ltd.,
D. INSECTICIDE + FUNGICIDE		
1.	Flubendiamide 3.5% + Hexaconzole 5% WG	M/s Rallis India Ltd.,
2.	Imidacloprid 18.5% + Hexaconzole 1.5% FS	M/s Rallis India Ltd.,

Total Formulation Registered : 736

LIST OF PESTICIDES WHICH ARE BANNED, REFUSED REGISTRATION AND RESTRICTED IN USE:

(As on 31.10.2019)

I. PESTICIDES / FORMULATIONS BANNED IN INDIA

A.	**Pesticides Banned for manufacture, import and use.**	
	1.	Aldicarb (vide S.O. 682 (E) dated 17th July 2001)
	2.	Aldrin
	3.	Benzene Hexachloride
	4.	Benomyl (vide S.O 3951(E) dated 8th August, 2018)
	5.	Calcium Cyanide
	6.	Carbaryl (vide S.O 3951(E) dated 8th August, 2018)
	7.	Chlorbenzilate (vide S.O. 682 (E) dated 17th July 2001)
	8.	Chlordane
	9.	Chlorofenvinphos
	10.	Copper Acetoarsenite
	11.	Diazinon (vide S.O 3951(E) dated 8th August, 2018)
	12.	Dibromochloropropane (DBCP) (vide S.O. 569 (E) dated 25th July 1989)
	13.	Dieldrin (vide S.O. 682 (E) dated 17th July 2001)
	14.	Endosulfron (vide ad-Interim order of the Supreme Court of India in the Writ Petition (Civil) No. 213 of 2011 dated 13th May, 2011 and finally disposed of dated 10th January, 2017)
	15.	Endrin
	16.	Ethyl Mercury Chloride
	17.	Ethyl Parathion
	18.	Ethylene Dibromide (EDB) (vide S.O. 682 (E) dated 17th July 2001)
	19.	Fenarimol (vide S.O 3951(E) dated 8th August, 2018)
	20.	Fenthion (vide S.O 3951(E) dated 8th August, 2018)
	21.	Heptachlor
	22.	Lindane (Gamma-HCH)
	23.	Linuron (vide S.O 3951(E) dated 8th August, 2018)
	24.	Maleic Hydrazide (vide S.O. 682 (E) dated 17th July 2001)

	25.	Menazon
	26.	Methoxy Ethyl Mercury Chloride (vide S.O 3951(E) dated 8th August, 2018)
	27.	Methyl Parathion (vide S.O 3951(E) dated 8th August, 2018)
	28.	Metoxuron
	29.	Nitrofen
	30.	Paraquat Dimethyl Sulphate
	31.	Pentachloro Nitrobenzene (PCNB) (vide S.O. 569 (E) dated 25th July 1989)
	32.	Pentachlorophenol
	33.	Phenyl Mercury Acetate
	34.	Sodium Cyanide (banned for Insecticidal purpose only vide S.O 3951(E) dated 8th August, 2018)*
	35.	Sodium Methane Arsonate
	36.	Tetradifon
	37	Thiometon (vide S.O 3951(E) dated 8th August, 2018)
	38.	Toxaphene(Camphechlor) (vide S.O. 569 (E) dated 25th July 1989)
	39.	Tridemorph (vide S.O 3951(E) dated 8th August, 2018)
	40.	Trichloro acetic acid (TCA) (vide S.O. 682 (E) dated 17th July 2001)
B.	**Pesticide formulations banned for import, manufacture and use**	
	1.	Carbofuron 50% SP (vide S.O. 678 (E) dated 17th July 2001)
	2.	Methomyl 12.5% L
	3.	Methomyl 24% formulation
	4.	Phosphamidon 85% SL
C.	**Pesticide / Pesticide formulations banned for use but continued to manufacture for export**	
	1.	Captafol 80% Powder (vide S.O. 679 (E) dated 17th July 2001)
	2.	Nicotin Sulfate
D.	**Pesticides Withdrawn** (Withdrawal may become inoperative as soon as required complete data as per the guidelines is generated and submitted by the Pesticides Industry to the Government and accepted by the Registration Committee. (S.O 915(E) dated 15th Jun,2006)	

	1.	Dalapon
	2.	Ferbam
	3.	Formothion
	4.	Nickel Chloride
	5.	Paradichlorobenzene (PDCB)
	6.	Simazine
	7.	Sirmate (S.O. 2485 (E) dated 24th September 2014)
	8.	Warfarin (vide S.O. 915 (E) dated 15th June 2006)

* Regulation to be continued in the extant manner for non-insecticidal uses.

II. PESTICIDES REFUSED REGISTRATION

S.No.	Name of Pesticides
1.	2,4, 5-T
2.	Ammonium Sulphamate
3.	Azinphos Ethyl
4.	Azinphos Methyl
5.	Binapacryl
6.	Calcium Arsenate
7.	Carbophenothion
8.	Chinomethionate (Morestan)
9.	Dicrotophos
10.	EPN
11.	Fentin Acetate
12.	Fentin Hydroxide
13.	Lead Arsenate
14.	Leptophos (Phosvel)
15.	Mephosfolan
16.	Mevinphos (Phosdrin)
17.	Thiodemeton / Disulfoton
18.	Vamidothion

III. PESTICIDES RESTRICTED FOR USE IN THE COUNTRY

S.No.	Name of Pesticides	Details of Restrictions
1.	Aluminium Phosphide	The Pest Control Operations with Aluminium Phosphide may be undertaken only by Govt./Govt. undertakings / Govt. Organizations / pest control operators under the strict supervision of Govt. Experts or experts whose expertise is approved by the Plant Protection Advisor to Govt. of India except [1]Aluminium Phosphide 15% 12g tablet and [2]Aluminum Phosphide 6% tablet. [RC decision circular F No. 14-11(2)-CIR-II (Vol. II) dated 21-09-1984 and G.S.R. 371(E) dated 20th may 1999]. [1] Decision of 282nd RC held on 02-11-2007 and, [2] Decision of 326th RC held on 15-02-2012. The production, marketing and use of Aluminium Phosphide tube packs with a capacity of 10 and 20 tablets of 3 g each of Aluminium Phosphide are banned completely. (S.O. 677 (E) dated 17th July, 2001)
2.	Captafol	The use of Captafol as foliar spray is banned. Captafol shall be used only as seed dresser. (S.O.569 (E) dated 25thJuly, 1989) The manufacture of Captafol 80 % powder for dry seed treatment (DS) is banned for use in the country except manufacture for export. (S.O.679 (E) dated 17th July, 2001)
3.	Cypermethrin	Cypermethrin 3 % Smoke Generator is to be used only through Pest Control Operators and not allowed to be used by the General Public. [Order of Hon'ble High Court of Delhi in WP(C) 10052 of 2009 dated 1407-2009 and LPA-429/2009 dated 08-09-2009]
4.	Dazomet	The use of Dazomet is not permitted on Tea. (S.O.3006 (E) dated 31st Dec, 2008)

S.No.	Name of Pesticides	Details of Restrictions
5.	Dichloro Diphenyl Trichloroethane (DDT)	The use of DDT for the domestic Public Health Programme is restricted up to 10,000 Metric Tonnes per annum, except in case of any major outbreak of epidemic. M/s Hindustan Insecticides Ltd., the sole manufacturer of DDT in the country may manufactureDDT for export to other countries for use in vector control for public health purpose. The export of DDT to Parties and State non-Parties shall be strictly in accordance with the paragraph 2(b) article 3 of the Stockholm Convention on Persistent Organic Pollutants (POPs). (S.O. 295 (E) dated 8th March, 2006)
		Use of DDT in Agriculture is withdrawn. In very special circumstances warranting the use of DDT for plant protection work, the state or central Govt. may purchase it directly from M/s Hindustan Insecticides Ltd. to be used under expert Governmental supervision. (S.O.378 (E) dated 26thMay, 1989)
6.	Fenitrothion	The use of Fenitrothion is banned in Agriculture except for locust control in scheduled desert area and public health. (S.O.706 (E) dated 03rdMay, 2007)
7.	Methyl Bromide	Methyl Bromide may be used only by Govt./ Government undertakings / Govt. Organizations / Pest control operators under the strict supervision of Govt. Experts or Experts whose expertise is approved by the Plant Protection Advisor to Govt. of India. [G.S.R.371 (E) dated 20thMay, 1999 and earlier RC decision]
8.	Monocrotophos	Monocrotophos is banned for use on vegetables. (S.O.1482 (E) dated 10th Oct, 2005)

S.No.	Name of Pesticides	Details of Restrictions
9.	Trifluralin	(vide S.O 3951(E) dated 8th August, 2018) (i) The Registration, import, manufacture, formulation, transport, sell and its all uses except use in wheat shall be prohibited and completely banned from date of publication of this Order. (ii) A cautionary statement has to be incorporated in the label and leaflet that it is toxic to aquatic organism, hence should not be used near water bodies, aquaculture or pisciculture area.

IV. PESTICIDES WHICH SHALL BE PHASED OUT VIDE GAZETTE NOTIFICATION NO. S.O. 3951 (E).

S.No.	Name of the pesticide	Insecticides to be phase out by 31st December, 2020
1.	Alachlor	(i) No new certificate of registration to manufacture shall be issued after publication of this Order. (ii) No person shall import, manufacture or formulate Alachlor with effect from the 1st January, 2019. (iii) The use of Alachlor shall be completely banned with effect from the 31st December, 2020. (iv) It is toxic to aquatic organism, hence a cautionary statement should be incorporatedon label and leaflets " toxic to aquatic organism hence should not be used near water bodies, aquaculture or pisciculture area.
2.	Dichlorovos	(i) No new certificate of registration to manufacture shall be issued after publication of this Order.

S.No.	Name of the pesticide	Insecticides to be phase out by 31st December, 2020
		(ii) No person shall import, manufacture or formulate dichlorvos with effect from the January, 2019. (iii) The use of dichlorvos shall be completely banned with effect from the 31st December, 2020. (iv) It is very toxic to aquatic organism, hence a cautionary statement should be incorporated on label and leaflets that it is toxic to aquatic organism, hence should not be used near water bodies, aquaculture or pisciculture area. (v) A warning may be incorporated in the label and leaflet stating that this product is toxic to honey bees so do not spray during active honey bees foraging period of the day.
3.	Phorate	(i) No new certificate of registration to manufacture shall be issued after publication of this Order. (ii) No person shall import, manufacture or formulate Phorate with effect from the 1st January, 2019. (iii) The use of Phorate shall be completely banned with effect from the 31st December, 2020. (iv) It is very toxic to aquatic organism, hence a cautionary statement should be incorporated on label and leaflets that it is toxic to aquatic organism, hence should not be used near water bodies, aquaculture or pisciculture area. (v) A warning may be incorporated in the label and leaflet stating that this product is toxic to honey bees so do not spray during active honey bees foraging period of the day.

S.No.	Name of the pesticide	Insecticides to be phase out by 31st December, 2020
		(vi) A cautionary statement should incorporate in label and leaflet that this product is toxic to birds. (iv) It is very toxic to aquatic organism, hence a cautionary statement should be incorporated on label and leaflets that it is toxic to aquatic organism, hence should not be used near water bodies, aquaculture or pisciculture area. (v) A warning may be incorporated in the label and leaflet stating that this product is toxic to honey bees so do not spray during active honey bees foraging period of the day. (vi) A cautionary statement should incorporate in label and leaflet that this product is toxic to birds.
4.	Phosphamidon	(i) No new certificate of registration to manufacture shall be issued after publication of this Order. (ii) No person shall import, manufacture or formulate Phosphamidon with effect from the 1st January, 2019. (iii) The use of Phosphamidon shall be completely banned with effect from the 31st December, 2020.
5.	Triazophos	(i) No new certificate of registration to manufacture shall be issued after publication of this Order. (ii) No person shall import, manufacture or formulate Triazophos with effect from the 1 st January, 2019. (iii) The use of Triazophos shall be completely banned with effect from the 31st December, 2020.

S.No.	Name of the pesticide	Insecticides to be phase out by 31st December, 2020
		(iv) It is very toxic to aquatic organism, hence a cautionary statement should be incorporated on label and leaflets that it is toxic to aquatic organism, hence should not be used near water bodies, aquaculture or pisciculture area. (v) A warning may be incorporated in the label and leaflet stating that this product is toxic to honey bees so do not spray during active honey bees foraging period of the day (vi) A cautionary statement should incorporate in label and leaflet that this product is toxic to birds
6.	Trichlorfon	(i) No new certificate of registration to manufacture shall be issued after publication of this Order. (ii) No person shall import, manufacture or formulate Trichlorfon with effect from the 1 st January, 2019. (iii) The use Trichlorfon shall be completely banned with effect from the 31st December, 2020. (iv) It is very toxic to aquatic organism, hence a cautionary statement should be incorporated on label and leaflets that it is toxic to aquatic organism, hence should not be used near water bodies, aquaculture or pisciculture area. (v) A cautionary statement should incorporate in label and leaflet that this product is toxic to birds.

Maximum Residue Limits of Fungicides Compiled by P. Dureja and VT Gajbhiye, IARI, New Delhi (Source: Pesticide Res. J. 16(2): 84–87)

S. No.	Name of fungicide	Food	Maximum residue limit (mg kg^{-1})
1.	Benomyl	Food grains	0.5
		Milled food grains	0.12
		Vegetables	0.50
		Mango	2.0
		Banana (whole)	1.0
		Other fruits	5.0
		Cotton seeds	0.10
		Groundnut	0.10
		Sugarbeet	0.10
		Dry fruits	0.10
		Eggs	0.10
		Meat and Poultry	0.10
		Milk and milk products	0.10
2.	Bitertanol	Wheat	0.05
		Groundnut	0.10
3.	Captafol	Tomato	5.00
4.	Captan	Fruits and Vegetables	15.00
5.	Carbendazim	Food grains	0.5
		Milled food grains	0.12
		Vegetables	0.50
		Mango	2.0
		Banana (whole)	1.0
		Other fruits	5.0
		Cotton seeds	0.10
		Groundnut	0.10
		Sugar beet	0.10
		Dry fruits	0.10
		Eggs	0.1
		Meat and Poultry	0.1
		Milk and milk products	0.1

S. No.	Name of fungicide	Food	Maximum residue limit (mg kg^{-1})
6.	Chlorothalonil	Groundnuts	0.10
		Potato	0.10
7.	Copper oxychloride	Fruits	1.0
		Potatoes	20.0
		Other vegetables	8.0
8.	Dithiocarbamates	Food grains	0.2
		Milled food grains	0.05
		Potatoes	0.1
		Tomatoes	3.0
		Cherries	1.0
		Other fruits	3.0
9.	Dodine	Apple	5.0
10.	Metalaxyl	Bajra	0.05
		Maize	0.05
		Sorghum	0.05
11.	Thiophanatemethyl	Apple	5.0
		Papaya	7.0
12.	Fenarimol	Apple	5.0
13.	Hexaconazole	Apple	0.1
14	Iprodione	Rape seed	0.5
		Mustard seed	0.5
		Rice	10.0
		Tomato	5.0
		Grapes	10.0
15	Tridemorph	Wheat	0.1
		Grapes	0.5
		Mango	0.05
16.	Penconazole	Grapes	0.2
17.	Propiconazole	Wheat	0.05
18.	Myclobutanil	Groundnut seed	0.1
		Grapes	1.0

S. No.	Name of fungicide	Food	Maximum residue limit (mg kg^{-1})
19.	Tricyclazole	Rice	0.01
20.	Dithianon	Apple	0.1
21.	Kitazin	Rice	0.2
22.	Cymoxanil	Grapes	0.1
23.	Triadimefon	Wheat Pea Grapes	0.5 0.1 2.0
24.	Fosetyl-Al	Grapes Cardamom	10 0.2

The principal rules were published vide number S.R.O. 2106, dated the 12th September, 1955 and were last amended vide G.S.R. 451 (E) dated 15th July, 2004.

16

SAFE HANDLING OF PESTICIDES

The following precautions should be strictly adhered to while handling pesticides:

1. Always store the products away from the reach of children, domestic animals and eatable articles.
2. Never store agrochemicals near any medicine.
3. Label the container properly.
4. Buy only recognized genuine packages of insecticides or fungicides and never transfer the concentrated product into another container.
5. Read carefully and follow strictly the instructions and precautionary measures sent with the container.
6. Use always the prescribed dosage and only for the purpose recommended. Never use chemicals for the control of flies, bugs etc. indoors or for riding animals from insects attacking them unless recommended.
7. Use rubber gloves when handling the concentrated product.
8. Do not mix the spray suspension with your hands, but use instead a rod or stick.
9. Do not go for any eatables while spraying or preparing the spray suspension.
10. After spraying, wash the hand properly with soap, preferably take bath and change the cloths.
11. Engage only adults and healthy workers and never children for spraying work.
12. Do not spray at a stretch. Spray only during the morning or late afternoon.

13. Pilots should never flu through their own spray mist or cloud.
14. In aerial operations, pilots should not help ground crew in loading pesticides in the aircraft.
15. Never blow into the clogged nozzles by mouth, but clean them with the help of a thin wire.
16. Always use sprayers and dusters in good condition. Leaky equipments may cause hazards to the worker.
17. Always fix a signboard to warn people that poisonous substance has been sprayed or dusted.
18. In case of uneasiness while working (vomiting sensation, intestinal cramps, giddiness etc.) call a doctor at once.
19. Do not spray or dust in tanks, ponds etc., which are sources of drinking water, supply or where fish are stocked.
20. Dispose by scattering over barren band or by burying the wash water of used equipments.
21. Empty packages (bottles, cans, tins etc.) should never be used for other purposes and are best destroyed or by breaking and buried deep in the ground.
22. Any person who as a result of any organo-phosphorus compound poisoning is under medical treatment should not work with chemicals such as Parathion, Rogor etc.

SELECTED REFERENCES

Agrios, G.N. 2005. *Plant Pathology*. Academic Press INC London, 803pp.

Butani, D. and Jotwani, MG. 2013. Insects in Vegetables. *Daya Publishing House*. 356pp.

David, M., Andrew, B., Carig, C. and Carol, M. 2003. Managing carrot rust fly: In search of alternatives for a tough customer. *Agrichemical & Environmental News*. 203: 1-9.

Gupta, P. R., Babu, B. R.R. M., Reddy, P. P., Kumar, N.K. K. and Verghese, A. 1998. Management ofHelicoverpa armigera on tomato with Trichogramma pretiosum and *Bacillus thuringiensis* var.*kurstaki. Advances in IPM for horticultural crops. Proceedings of the first national symposium on Pest management in horticultural crops: environmental implications and thrusts, Bangalore, India*. p. 75-80

Gupta, S.K. and Handa, Anil 2009. *Ecofriendly Management of Diseases of Pea and otherVegetables* . Dept. of Mycology and Plant Pathology, Dr. Y.S. Parmar University of Horticulture & Forestry, Nauni, Solan, 101pp.

Gupta, S.K. and Thind, T.S. 2018. *Disease Problems in Vegetable Production 2nd Edition*.Scientific Publishers (India), Jodhpur, 586pp.

Gupta, S.K., Sharma, R.C. and Sharma, Monica 2017. *Diseases of Vegetable, Ornamental and Spice Crops*. Scientific Publishers (India), Jodhpur, 193pp.

http://agricoop.nic.in/sites/default/files/ICAR_7.pdf

http://agropedia.iitk.ac.in/content/tomato-pinworm-tuta-absoluta-new-invasive-pest-india

http://ppqs.gov.in/divisions/cib-rc/registered-products

http://www.researchjournal.co.in/online/RKE/RK%20Eng%20%20%2010%20(2)/10_25-27.pdf

https://www.academia.edu/10292243/Insect_Pests_of_Cruciferous_Vegetables_and_Their_Management

https://www.arbico-organics.com/category/pest-solver-guide-whitefly

https://www.researchgate.net/publication/275036500_First_report_of_Tomato_leaf_miner_Tuta_absoluta_from_India_httpwwwindianjournalscomijoraspxtargetijorijevolume77issue4article003

https://www.researchgate.net/publication/281454321_Monitoring_of_Helicoverpa_armigera_using_pheromone_traps_and_relationship_of_moth_activity_with_larval_infestation_on_Carnation_Dianthus_caryophyllus_in_Darjeeling_Hills

https://www.slideshare.net/ravindra43/major-insectpests-of-solanaceous-crop-tomato-potato-and-their-management

McKinlay, Roderick G. 2019. *Vegetable crops pests (Ed.).* Springer Nature Switzerland AG. Part of Springer Nature. 384p.

Rai, AB. *Integrated pest management for vegetable crops.* (ICAR) Crop Protection Division, Indian Institute of Vegetable Research Varanasi http://agricoop.nic.in/sites/default/files/ICAR 7.pdf

Singh, R.S. 1995. *Diseases of Vegetable Crops3rd Edition.* Oxford and IBH publication Pvt. Ltd. 325pp.

Verma, L.R. and Sharma, R.C. (eds.). 1999. *Diseases of Horticultural Crops- Vegetables, Ornamentals and Mushrooms.* Indus Publishing Co., New Delhi, 723pp.

GLOSSARY

Abdomen: In insects, the body is divided into three parts; head, thorax and abdomen. The third part of the body is called as abdomen.

Acervulus: A sub-epidermal saucer shaped, asexual fruiting body producing short conidiophores and conidia.

Actinomycete: Formerly, a group of bacteria forming branching filaments.

Aeciospore: A binucleate rust spore produced in an aecium.

Aecium: A cup shaped fruiting body of the rust fungi which produces aeciospores.

Antheridium: The male sexual organ found in some fungi.

Anthracnose: These are suppressed lesion resembling a leaf or pod spot which are caused by fungi that produce acervuli.

Antibiotic: A Chemical compound produced by one microorganism which inhibits or kills other microorganisms.

Apothecium: An open cup or saucer shaped ascocarp of some ascomycetes.

Appressorium: The Swollen tip off a hypha or germ tube that facilitates attachment and penetration of the host by the fungus.

Ascospore: A sexually produced spore borne in ascus.

Ascus: A sac like cell of a hypha containing ascospores (usually eight)

Asexual Reproduction: A mode of reproduction not involving the union of gametes or meiosis.

Atrophy: Inhibition of plant growth resulting in stunting or dwarfing.

Autoecious fungus: A parasitic fungus that can complete its entire life cycle on the same host.

Bacterium: A unicellular prokaryotic microscopic organism that lacks chlorophyll and multiplies by fission.

Basidiospore: A sexually produced spore borne on a basidium.

Basidium: A Club shaped structure on which basidiospores are borne.

Beetle: The adult insects of Order Coleoptera. The name Coleoptera, derived from the Greek words *"koleos"* meaning sheath and *"ptera"* meaning wings, refers to the modified front wings which serve as protective covers for the membranous hindwings. Thus beetles can be defined as an insect which have forewings that are modified into hard wing cases (elytra) which cover and protect the hindwings and abdomen.

Biodegradation: The Process whereby a compound is decomposed by natural biological activity. A biodegradable substance is one in which a microbe can reduce it to CO_2 and H_2O by mineralization.

Biological control: Biological control is a component of an integrated pest management strategy. It is defined as the method of controlling the pests (insects, mites, plant pathogens, weeds etc) by the use of predators, parasitoids, pathogens or any other living organisms to reduce the population of pest.

Bio-pesticide: These are the types of pesticides which are derived from natural materials such as animals, plants, bacteria. These are of two types: i) **Animal origin** eg. Predator, parasitoid, pathogen ii) **Plant origin** eg. plant based formulations or products like Neem oil.

Bioremediation: The use of biological agents to render hazardous wastes to non hazardous or less hazardous form.

Biotroph: An organism that can live and multiply only on other living organism.

Biotype: A subgroup within a species or race, usually characterized by the common possession of a single or a few new characters.

Bipectinate: *Bi*= Two; Latin *pectin,* the linguistic stem of *pecten,* a comb or toothed; resembling the teeth of a comb. Antennae having comb-like toothed structure on both sides.

Blight: A disease characterized by general and rapid killing of leaves, flowers and stem of plants.

Blister: A raised lesion on the leaf and stem surface which exposes spores when burst.

Blotch: A disease characterized by large and irregular spots or blotches on leaves, shoots and stems of a plant.

Blotching: A discoloration on leaves or fruits due to fungi, bacteria, mite etc

Botanical pesticide: These are the pesticides of plant origin.

Bronzing: It refers to the way some leaves or fruit turn purplish or bronze-colored as a result of mineral imbalances, pest feeding, chemicals, environmental conditions, or disease. Bronzed leaves tend to be smaller than normal and the damaged areas are unable to perform photosynthesis.

Bt. ***(Bacillus thuringiensis)*:** It is a spore forming bacterium that produces crystals proteins (cry proteins), which are toxic to many species of insects especially the Lepidopteran insect pests and is used as bio-pesticides.

Bug (True bug): The insects which belong to Order: Hemiptera. In this the forewings are leathery or parchment-like at the base and membranous near the tip and are hardened throughout the proximal two-thirds, while the distal portion is membranous.

Canker: A necrotic , often sunken lesion on a stem , branch, or twig of a plant.

Caterpillar: It is a type of polypod larva which bears 3 pairs of thoracic legs and 5 pairs of prolegs. The prolegs are present on 3rd, 4th, 5th, 6th, and 10th abdominal segments e.g. larva of Lemon butterfly, larva of gram pod borer etc.

Cauda: It is called as the tail of aphids and is used to excrete excess of sugar from the body in the form of droplets of honeydew.

Chelceral stylet: These are the mouth parts of organisms belonging to Class Arachnida (Mite, scorpion etc.). The chelae are evolved into sharp mouthparts that mites use to pierce the surface of the plants they feed on in order to suck out the contents of the plant cells.

Chemical control: When different types of chemicals (pesticides) are used to kill pests, it is called as chemical control.

Chlamydospore: A thick walled asexual resting spore formed by the modification of a cell of a fungus hypha.

Chlorosis: Yellowing of normally green tissue due to chlorophyll destruction or failure of chlorophyll formation.

Chrysopid: These are called as the green lacewings which belong to family Chrysopidae and Order Neuroptera.

Clavus: The oblong scelerite at the base of the inferior margin of the hemelytron in Heteroptera. (Sclerite: Hard plate).

Cleistothecium: An entirely closed ascocarp.

Cocoon: It is the third stage in the life cycle of some categories of insects belonging to Order: Lepidoptera. It is the silky case spun by the larvae for the protection in pupae.

Coenocytic: Mycelium without any septum resulting in multinucleate condition.

Combi-products: These are the combination of mixing together insecticides or fungicides or weedicides.

Concentric: Forming one circle around another with a common center.

Conidiophore: A specialized hypha on which one or more conidia are produced.

Conidium: An asexual fungal spore formed from the end of a conidiophore.

Corium: The corium is the thickened, leathery, basal portion of an insect forewing or hemelytron in the order Hemiptera. Specifically the inner cell of the basal portion is the corium.

Cornicles: These are small tubes that project from abdomen in aphids which release alarm pheromone.

Crinckling: The condition of leaves which is characterized by puckering, mottling and distortion of the leaves.

Cu2-vein: The second branch of cubitus in wings of some insects.

Cultural control: The agronomical practices of modifying the growing environment to reduce the abundance of insect pests and diseases.

Curling: The condition of plants which shows the twisting (curling) of leaves.

Damping-off: Destruction of seedlings near the soil line, resulting in the toppling over of seedlings on the ground.

Die-back: Disease in which progressive death of shoots and branches occur from tip to backwards.

Disease cycle: The chain of events involved in disease development, including the stages of development of the pathogen and the effect of the disease on the host.

Disease: Disease is a malfunctioning process that is caused by continuous irritations which results in some suffering producing symptoms.

Disorder: Non-Infectious harmful deviation due to abiotic factors in plants e.g. mineral deficiency or toxicity, adverse soil temperature and soil moisture conditions.

Dissemination: Transfer of inoculums from its source to healthy plants.

Downy mildew: A plant disease in which the mycelium and spores of the fungus appear as a downy growth on the under surface of the host caused by fungi in the family Peronosporaceae.

Entomophatogenic: These are the microorganisms which are pathogenic to insects.

Epidemic: A widespread and severe outbreak of a disease.

Epidemiology: The study of factors affecting the outbreak and spread of infectious disease.

Epiphytotic: A widespread and destructive outbreak of a disease of plants.

Eradication: Control of plant disease by eliminating the pathogen after it is established or by eliminating the plant that carry the pathogen.

Feign death: It is the false appearance in which the insect pretends itself to be dead.

Ferruginous red: When the leaves show iron rust or reddish-brown colour.

Flagellum: A whip like structure projecting from a bacterium or zoospore and functioning as an organ of locomotion.

Fleck: A minute spot.

Formulation: It is the mixing of active ingredient (insecticides/ fungicides/ weedicides) and inert material to form a product which improve the storage, handling, safety, application, or effectiveness of any pesticide.

Fruiting body: A complex fungal structure containing spores.

Fungicide: A chemical compound toxic to fungi that kills or inhibit the growth of the fungi.

Fungus: Undifferentiated plant lacking chlorophyll and conductive tissues.

Fusiform: This is the shape which is found in some insect which tapers at both ends and is spindle-shaped.

Gall: A swelling, or overgrowth produced on a plant as a result of infection by certain pathogens.

Gene flow: the exchange of genetic information among individuals, populations, or species. With the preservation of that genetic information in the following generations.

Gene silencing: partial or total suppression of gene expression.

Genotype: the aggregate of genes in an organism.

Gnaw: The condition in which some insects bite at or nibble the plant parts.

Gregarious: When insects or any of their stages are found in groups.

Grub: The larval stage of order Coleoptera.

Haustorium: A projection of hyphae into host cells which acts as a penetration and absorbing organ.

Heteroecious: Requiring two different kinds of hosts to complete its life cycle, pertaining particularly to rust fungi.

Hopper burn: It is the condition of leaves distinguished by yellowing, burning and subsequently burning of plants due to excessive feeding of hopper.

Host range: The various kinds of host plants that may be attacked by a parasite.

Host: A plant that is invaded by a parasite and from which the parasite obtain its nutrients.

Hyaline: colorless , transparent.

Hybrid: The offspring of two individual different in one or more heritable characteristics.

Hybridization: The crossing of two individuals differing in one or more heritable characteristics.

Hydathodes: Structure with one or more openings that discharge water from the interior of the leaf to its surface.

Hyperplasia: A plant overgrowth due to increased cell division.

Hypersensitivity: Excessive sensitivity of plant tissues to certain pathogens. Affected cells are killed quickly, blocking and advance of obligates parasites.

Hypertrophy: A plant overgrowth due to abnormal cell enlargement.

Hypoplasia: Same as atrophy.

Immune: Exempt from infection by a given pathogen.

Imperfect fungus: A fungus that is not known to produce sexual spores.

Incubation period: The time lapse between invasion of a host by a pathogen and appearance of symptoms on the host.

Infection: The establishment of a host parasitic relationship within a host system.

Inoculum: The pathogen or its parts that can cause disease. The portion of individual pathogens that are brought into contact with the host.

Insecticides: Any chemical or its product used to kill insects.

Integrated control: An approach that attempts to use all available methods of control of a disease or of all the diseases and pests of a crop plant for best control results but with the least cost and the least damage to the environment.

Kerosinized water: Mixing of kerosene with water.

Lacerating the leaf: Scratching of leaves due to infestation of thrips.

Lacewing: Insects of Families Chrysopidae and Hemerobiidae of Order Neuroptera with gauzy wings which feed on aphids.

Larva: Immature form of an insect which comes out immediately after hatching of eggs in holometablous insects.

Leaf spot: A self-limiting lesion on a leaf.

Lesion: A localized area of discolored, diseased tissue.

Life cycle: The stage or successive stages in the growth and development of an organism that occur between the appearance and reappearance of the same stage (e.g spore) of the organism.

Light trap: It is a device meant for collecting or destroying insects. It consists of a bright light and a trapping or killing medium.

Looper: It is a type of polypod larva which bears 3 pairs of thoracic legs and two pairs of prolegs on 6th and 10th abdominal segments e.g. cabbage looper.

Manual collection: Collection of insects manually for scientific studies.

Mass trapping: A method of collection of insects in large numbers by the use of an attractant (pheromone or in some cases a specific colour) in order to reduce their numbers.

Mechanical control: It is the management of insect-pests by the use of mechanical force or manualy without harming the ecosystem.

Mildew: A fungal disease of plants in which the mycelium and spores of the fungus are seen as a whitish growth on the host surface.

Mine: The damage of leaves due to some Dipteran and Lepidopteran pests which is distinguished by making of small tunnels.

Mite: A small minute arachnid which has four pair of legs.

Moerick yellow water trap: It is the water trap used to trap aphids.

Mosaic: Symptom of certain viral diseases of plants characterized by intermingled patches of normal and light green or yellowish colour.

Mottle: An irregular pattern of indistinct light and dark areas.

Mould: Any profuse or wooly fungus growth on damp or decaying matter or on surface of plant tissue.

Mycelium: The hypha or mass of hyphae that make up the body of a fungus.

Necrosis: it refers to death of a plant part usually of a clearly delimited part of a plant or of a tissue.

Neem seed kernel extract: The extract prepared by mixing neem seed kernel with water or any other solvent.

NPV (Nuclear Polyherdral Virus): These are polyhedral inclusion bodies used to kill some insect larvae (preferably the larvae of moth and butterfly).

Nymph: Immature form of paurometabolous insects.

Obligate parasite: A parasite that in nature can grow or multiply only on living organism.

Oogonium: The female gametangium of some oomycetes containing one or more gametes.

Oospore: A sexual spore produced by the union of two morphologically different gametangia.

Opalescent: It is yellowish-red colour in transmitted light and blue in the scattered light perpendicular to the transmitted light.

Oval: It is rounded and slightly elongated shape just like that of an egg.

Palisade mesophyll tissue: The plant cells which collectively make up the mesophyll tissue in plant leaves. This layer (palisade layer) is located beneath the upper epidermis of the leaf.

Parasite: An organism living on or in another living organism and obtaining its food from the latter.

Parasitoid: An insect which kills any of the stage of the other insect host and is free living in its adult stage.

Pathogen: A causal agent or an organism or virus able to cause disease in a particular host or range of hosts.

Pathogenesis: The sequence of processes in disease development from the initial contact of a pathogen with its host for the completion of its syndrome.

Pathogenicity: The relative capability of a pathogen to cause disease. The term is frequently given quantitative qualification(high, moderate and low) and used synonymously with virulence.

Perfect stage: The sexual stage (fruiting bodies) in the cycle of a fungus.

Perithecium: The globular or flask shaped ascocarp of the pyrenomycetes, having an opening or pore (ostiole).

Pesticides: Any chemical or its product which is used to kill pests.

Pheromone: These are exocrine secretions released by an insect into the environment which affects the behavior and physiology of other insects of the same species.

Phloem: Food conducting tissue, consisting of sieve tubes, companion cells, phloem parenchyma and fibers.

Phytoplasma: mollicutes that infect plants and can not yet be grown in culture, as contrasted to Spiroplasmas, which can be cultured.

Phytseiid mite: These are those mites which are used as a biological control agent for managing mite pests and thrips.

Plasmodium: A naked, slimy mass of protoplasm containing numerous nuclei.

Predator: An animal (insect) that naturally preys on others (insects) and usually kills it.

Primary infection: The first infection of a plant in the spring by the over wintering pathogen.

Primary Inoculum: The over wintering pathogen or its spores that cause primary infections.

Protectant: A substance that protects an organism against infection by a pathogen.

Proteomics: The science of identifying the sequence, function, structure, and interrelationship of all proteins in an organism and subsequent processing of that information.

Pupa: It is the inactive immature form between larva and adult in some insects (holometabolous insects)

Pustule: Small blister like elevation of epidermis created as spores form underneath and push outward.

Pycnidium: An asexual, spherical, or flask shaped fruiting body lined inside with conidiophores and producing conidia.

Pycniospore: Also called as spermatium. A spore produced in a pycnium.

Pycnium: Also called as spermatium. A fruiting body of the rust fungi that produces small spores called as pycniospores or spermatia which cannot infect plants but function as gametes or gametangia.

Pyriform: The pear-shape assumed by an organism.

Quarantine: Legislative measure for the control of import and export of plants or other propagating material to prevent spread of disease and pests from one location to another.

Race: A genetically and often geographically distinct mating group within a species, also a group of pathogens that infect a given set of plant varieties.

Resistance: The ability of the host to overcome completely or to some degree the effect of a pathogen or other damaging factor.

Resting spore: A sexual or other thick walled spore of a fungus that is resistant to extreme in temperature and moisture and which often germinates only after a period of time from its formation.

Rot: The softening, discolouration and often disintegration of a succulent plant tissue as a result of fungal or bacterial infection.

Rust: Reddish or rust coloured pustules of spores which give rusty appearance to the infected portion of plant.

Sanitation: The removal and burning of infected plant parts, decontamination of tools, equipment hands etc.

Saprophyte: An organism that uses dead organic matter for food.

Sclerotium: A hard resting body resistant to unfavorable conditions which may remain in soil for long periods.

Scutellum: It is the posterior portion of either the mesonotum or the metanotum of an insect thorax

Scutum: A shield-shaped dorsal plate in some insects as in grasshopper.

Secondary infection: Any infection caused by inoculums produced as a result of a primary or a subsequent infection, an infection caused by secondary inoculums.

Secondary inoculums: Inoculum produced by infections that took place during the same growing season.

Semilooper: It is a type of polypod larva which bears 3 pairs of thoracic legs and 3 pairs of prolegs. Prolegs are present on 5, 6, and 10th abdominal segments e.g. Castor Semilooper, Cotton Semilooper etc.

Septate: Having cross walls.

Septum: A cross wall (in hypha or spore).

Setaceous: Consisting of seta or bristle.

Sex pheromone: These are exocrine secretions released by either sex to attract the opposite sex.

Sexual dimorphism: It is the difference in appearance between male and female.

Shot hole: A symptom in which small diseased fragments of leaves fall off and leave small holes in their place.

Sign: The expression of a pathogen or its parts on a host.

Silken web: When some insects use silk to make web for their protection.

Silvery white sheen: Showing white lustrous surface.

Skeletonized: When green portion of leaf is eaten by an insect leaving only veins and mid rib.

Sky blue sticky trap: These traps are used for monitoring of thrips, as they are particularly attracted to this colour.

Smut: A disease characterized by the sooty mass of black or purplish black powdery spores usually appearing on floral organs.

Solitary bees: Those bees which are not social and live alone. They do not produce honey and do not live in hives.

Sorus: A compact mass of spores of fruiting structure found especially in the rusts and smuts.

Spermatium: The male gamet or gametangium of the rust fungi.

Spindle shaped: Insects having circular cross section like bodies and tapering towards each end.

Sporangiophore: A specialized hypha bearing one or more sporangia.

Sporangium: A container or case of asexual spores.

Spore: The reproductive unit of fungi consisting of one or more cell, it is analogous to the seed of green plants but without an embryo.

Sporodochium: A fruiting structure consisting of a cluster of conidiophores woven together on a mass of hyphae.

Sporulate: To produce spores.

Sticky tarp: These are glue-based traps which are used for monitoring and mass trapping of small insects.

Strain: The descendants of a single isolation in pure culture, an isolate, also a group of similar isolates, a race .in plant viruses, a group of virus isolates having most of their antigens in common.

Susceptibility: The inability of a plant to resist the effect of a pathogen or other damaging factor.

Symptomatology: The science concerning study of disease symptoms of a plant.

Symptoms: The external and internal expression or alteration of a plant as a result of the disease.

Syndrome: The totality of effects produced in a plant by a disease, whether simultaneously or successively including those effects not directly detectable to the unaided eye.

Syrphid: These are flies of Order Diptera. They are efficient pollinators and some members of these are predators in their larval stage.

Teliospore: The sexual, thick walled resting spore of the rust and smut fungi.

Termite: These are small, soft-bodied insects which live in colony and have different castes.

Thorax: Thorax is the second main division of an adult insect.

***Trichogramma* sp.:** These are small polyphagous wasps, many of which live as endo-parasitoids on the eggs of other insects.

Twisting: When the leaves take a spiral form.

Upward cupping: It is the upward curling of leaves.

Urediniospore: A binucleate, repeating spore of the rust fungi.

Uredium: The fruiting structure of rust fungi in which urediniospores are produced.

Vascular: Term applied to a plant tissue or region consisting of conductive tissue, also to a pathogen that grows primarily in the conductive tissue of a plant.

Vector: An animal able to transmit a pathogen.

Vesicle: A bubble-like structure produced by a zoosporangium and in which zoospores are released or are differentiated.

Viroids: Small, low molecular weight ribonucleic acids (RNA) that can infect plant cells, replicate themselves, and cause disease.

Virus: A submicroscopic obligate parasite consisting of nucleic acid and protein.

Webbing: Resembling web or net like appearance on leaves.

Wilt: Loss of turgidity and drooping of plant parts due to non-availability of water.

Withering: When the leaves become dry and sapless.

Xylem: A plant tissue consisting of tracheid's, vessels, parenchyma cells, fibers and wood.

Zoosporangium: A sporangium which contains or produces zoospores.

Zoospore: A spore bearing flagella and capable of moving in water.

■